Track of the
COYOTE

NorthWord Press, Inc.
P.O. Box 1360
Minocqua, WI 54548

Cover design by Russell S. Kuepper
Book design by Lisa Moore

Library of Congress Cataloging-in-Publication Data
    Wilkinson, Todd.
        Track of the Coyote / by Todd Wilkinson; photography by Michael H. Francis.
            p.   cm.
        ISBN 1-55971-471-9
        1. Coyotes—Yellowstone National Park. 2. Coyotes—Behavior—United States.
    3. Coyotes—folklore. 4. Coyotes—Control—United States. 5. Indians of North
    America—Folklore. I. Francis, Michael H. (Michael Harlowe). II. Title.
    QL737.C22W544   1995
    599.74'442—dc20   95-15125

Printed in Hong Kong

Todd Wilkinson

Track of the
COYOTE

Photography by Michael H. Francis

NorthWord
PRESS, INC.

Minocqua, Wisconsin

DEDICATION

For Jeanne, and Carter, our own pup

AUTHOR'S ACKNOWLEDGMENTS

The author extends his sincere gratitude and admiration to Bob Crabtree, the canid biologist whose story is told in this book and the researcher who is trying to engender a new cultural appreciation for this amazing predator. Bob was generous with his time and knowledge in reviewing the manuscript and sharing new information; to Adolph and Olaus Murie who recognized the intrinsic value of coyotes long before it became fashionable; to Franz Camenzind, Paul Paquet, Jim Till, Dick Randall, George Horse Capture, Weldon Robinson, Bob Murphy, Jennifer Sheldon, Teddy Thompson and Durward Allen who each offered pieces of personal insight into the nature of the beast. To Barbara Harold, Managing Editor at NorthWord Press, who oversaw the completion of this project and made it a far better book as a result; to the late Robert Smith for enlivening my childhood with tales about Minnesota wolves and "brush wolves"; and finally, a special note of appreciation to John Varley, director of the Yellowstone Center for Research in Yellowstone National Park who had the courage to launch a major study of unexploited coyotes in the park in order to compile an ecological profile that previously did not exist. I hope the park is successful in keeping the Yellowstone coyote project going, for it would be a shame and a tragic scientific loss to not document the interaction between wolves and coyotes in the years ahead, especially after such important groundwork has been laid.

PHOTOGRAPHER'S ACKNOWLEDGMENTS

The photographer would like to thank the following individuals for their help with this project. My wife Victoria for her support and understanding, Gary Leppart, Ron Shade, Rona Johnson, Troy and Kirsten Hyde, Jay Diest, and especially Dr. Robert Crabtree.

CONTENTS

The coyote is the most persecuted predator in North America, and surprisingly one of the least understood. Only by taking the time to educate oneself about its habits is it possible to fully appreciate its important natural role.

IN THE BEGINNING, BEFORE THERE WAS EARTH OR SKY, darkness or sunlight, the Creator fashioned a mischievous four-legged animal out of clay and called it coyote. Above all other creatures in the universe, coyote alone was chosen to hold dominion over the world. Forevermore, it would teach its neighbors valuable lessons by serving as friend and prophet.

Fortunately, according to ancient Native American legends, the higher power possesses a passion for beautiful things, a keen sense of humor and a desire for sonorous music. That is why coyote was given the alluring shape of a dog, a brilliant mind to play tricks on his enemies, and vocal chords to fill the night with its rapturous song.

Even today, the music continues. As humankind prepares to enter a new millennium, coyote's presence in nature has hardly diminished. His numbers may even be growing.

There are many people who claim that this enterprising predator is the smartest land animal on the planet. Others attribute magical powers to the animal's yellowish stare. Indeed, coyote is one beast that lives up to its mythological reputation. He is there, watching us. If only we take the time to look, and listen, it is our ears that will be rewarded most, with a song.

Among all the sensations that greet us in nature, nothing rings quite so sweet as the a cappella howls of coyotes baying at the rising moon. This is the tribute they offer to the heavens. Wherever we travel on the continent, the sounds of coyotes are never very far away. Their yips and barks are reminders of the first canid greetings that resonated through human ears thousands of years ago.

And yet, for centuries, human folklore has judged the coyote to be inferior compared to its close relative, the wolf. The bias has always been that coyotes are somehow the lesser species because they don't seem to have the same social, or pack, structure. But when left to roam free without persecution, coyotes are far more complex and sophisticated than we have been led to believe. Contemporary scientific research is painting a fresh portrait of the coyote, and with it comes a new appreciation of the animal's role in wild ecosystems.

From Los Angeles to New York City, Americans have coyotes on their minds—and whether we choose to identify the animals as friends or foes, they are there to teach us about our environment. The Native American elders of long ago realized that the coyote was sent from the One Above to be a sign of Good Medicine. Now it is our turn to recognize the wild dog for what it is. Howling out there not far from where we live is a messenger called the singing trickster, and he's arrived to announce his presence as coyote, America's little wolf.

Nearly a dozen different types of distinct coyote vocalizations have been documented by researchers. Barks, yips, howls and growls are the foundation of a coyote's ability to communicate with others of its kind.

THE MATRIARCH OF LAMAR

SHE HAD BEEN THE ELUSIVE ONE, outwitting a team of scientific trappers who tried to predict her next step. Her intuition told her to stay away from the steel jaws buried in the dirt, to guard herself from the creatures who walked on two legs. While other members of her tribe were far less callow in their aversion to humans, she permitted the onlookers to watch her but only from a distance, because she knew what happens when you venture too close to Homo sapiens. In the ebbing hours of daylight, they caught a squinting glimpse of her and in the next second she was gone, yipping a lament from some hidden vantage.

Until this spring morning, the matriarch of Lamar Valley had managed to escape the official I.D. number that researchers were hoping to give her. In the eyes of her pursuers, before they realized her stature, she had been considered merely an anonymous female coyote leading her pack across a secluded valley in the maw of Yellowstone National Park. But Number 570,

A young coyote stops long enough from mousing to curiously view the photographer.

as she became known to biologist Robert Crabtree, represented no ordinary wild canid. An alpha female, she would take her place as the queen of the largest and most successful coyote pack inhabiting the northern Rocky Mountains. The life of her clan, the Bison Peak pack, provides a window for humans to peer inside the secret society of America's singing tricksters.

To understand the significance of the Lamar matriarch first requires the acquaintance of the person trying to catch her. Bob Crabtree, a tall, sturdy ecologist with a bushy dark mustache, was selected by the National Park Service to take a five-year look at the last major unexploited coyote population left in the lower 48 states. Unexploited, meaning coyotes subsisting in a natural environment without severe disruptions to either their habitat or their numbers by humans. Such places are hard to find anymore, particularly in a modern world which has mixed feelings about the proper place of ubiquitous canid predators.

Only a handful of geographical spots fit the parameters of isolation where large communities of coyotes still can be observed virtually free of human interference—chief among them is the sanctuary of Yellowstone Park which encompasses roughly 3,600 square miles; there is also a remote section of the Hanford nuclear compound called the Arid Lands Ecology Reserve (which sits within a super-guarded area in Washington State where plutonium is produced to make nuclear bombs); and a sweep of sagebrush enveloped by Grand Teton National Park and the National Elk Refuge in the shadow of the Tetons. Much of the body of existing scientific literature relating to non-persecuted coyotes stems directly from this trio of outposts.

Crabtree's assignment was historic and daunting for a couple of reasons. First, it came fifty years after a landmark study by naturalist Adolph Murie, who in 1940 published a seminal report entitled "Ecology of the Coyote in Yellowstone" that shed new light on the coyote's food habits. Second and most tantalizing, Crabtree was asked to chronicle the ways that coyotes had filled the vacant niche of wolves following their extirpation from the ecosystem in the 1920s.

Coyotes had emerged not only as the top dog in Yellowstone, but over the same time span they solidified their presence across North America in general. Ironically, his findings are revealing because, with wolves being reintroduced to the park in the 1990s, the information establishes a framework to study the interaction between these largest members of the wild dog family.

As a culture, we know a lot about how to kill coyotes and how to curse

Preceding page. When given the choice, coyotes prefer to see and not be seen.

Biologist Bob Crabtree was hired to spearhead a ground-breaking study into coy-ote behavior and its ecological role prior to the return of wolves in Yellowstone National Park. He hopes his research will help engender a new view of the maligned species. Here, Crabtree prepares a special trap to catch one of the famed Bison Peak pack members.

The coyote's sense of smell enables it to detect food sources and enemies miles away. On the Alaska tundra, a lone coyote sniffs the wind, always alert to the possibility of nearby wolves.

them, but we seldom learn anything about how to get along with them or appreciate what makes them tick. Coyotes are showing up in our cities, suburbs and rural environs from the Atlantic to the Pacific. Everywhere they go, they are turning heads because it is rare indeed to have these efficient large carnivores residing near our doorsteps.

While coyotes have become icons of folk culture in some circles, they still are vilified as Public Enemy Number One with a bounty on their heads in rural America. Almost everyone you talk to has a different impression of coyotes and indeed, these dogs are a bundle of contradictions—soft and gentle, social and fiercely loyal to their own family values; sharp-toothed and vicious and capable of killing for their next meal. All of these traits, not unlike those in humans, are rolled up into the character that Native Americans call the mythical "trickster."

"Coyotes are the most effective large predator in America, but few people realize it," Bob Crabtree said one day as we drove to Yellowstone in his pick-up truck, looking for the elusive matriarch of Lamar. "There are many theories about the role of predators in ecological communities. Here's our best chance to look at how it all works."

Yellowstone remains a novel laboratory for watching how canids function as predators without being preyed upon by people. A century ago, *Canis latrans* was a creature limited chiefly to plains and desert environs west of the Mississippi River. But these kindred cousins of the wolf and common dog have proven themselves to be masters of adaptation. Today, the coyote is the most widely distributed, studied and persecuted wild dog species on the continent.

Crabtree's intention was to go back in time and witness the highly-evolved and proper behavioral complexities of animal behavior to help paint a new picture of the beast. Begun in the early part of the 1990s, his project is the most ambitious ever launched to radio collar and track what many have called the "little brother of the wolf."

In Yellowstone, Crabtree felt he had big shoes to fill. What was intimidating was Murie's work (and a few other pioneering studies that followed). Murie was a scientist who was well ahead of his time. He respected the coyote and championed the value of predators in ecosystems. Because of Adolph Murie public attitudes changed.

Despite its reputation, Yellowstone was not always insulated from predator control campaigns that tinkered with the mechanisms of the food chain. When the park was set aside by Congress in 1872, sportsmen (who then hunted in the

A mated alpha pair concentrate on their approach toward unsuspecting prey.

Preceding page. By engaging in mock battles, pups learn important lessons in defense.

park) wanted predators destroyed because they posed an alleged threat to the health of big game herds. Ranchers also sought to eliminate all meat-eating competitors, fearing they might snack on their cattle and sheep.

So the federal government stepped in and acted on their behalf by targeting the biggest of the carnivores first—grizzly bears, mountain lions, and wolves—leaving coyotes in the last half of this century to receive the full brunt of anti-predator sentiments, sentiments that still linger and cost taxpayers more than $30 million a year. What is the coyote's future? Perhaps the best place to begin answering the question is to explore the interior of the world's oldest national park.

THE TERRITORY

Loosely delineated across the Lamar Valley and Blacktail Plateau are the territories for twenty-four distinct coyote packs. Were you to stretch the scope of observation a dozen miles in either direction so that it included the entire northern tier of Yellowstone from Mammoth Hot Springs (park headquarters) to the small town of Cooke City at the park's northeastern gate, there would be nearly 500 coyotes represented in sixty packs.

Supreme among them, the Bison Peak pack holds dominion over the best chunk of coyote real estate Crabtree had ever seen. The northern perimeter of the territory is bounded by the foot of Bison Peak itself; the southern extension by riparian swales along the Lamar River; the east flank borders with the Druid Peak pack at the historic Buffalo Ranch Ranger Station; and the west extends toward the mouth of Lamar Canyon inhabited by the namesake Lamar Canyon pack.

"As an observer of coyotes, my favorite seasons are the spring and the fall," Crabtree says as we wade through ankle-deep fescues covered by spring dewdrops. It is the coyote pupping season, the weeks when coyotes produce young to sustain the lifeblood of the pack. "These are the months of the year when you learn the most about the things you are trying to study. Among these coyotes you see a cohesive bond of social interaction. Right now we're standing in the middle of exceptional coyote range. This is as good as it gets."

Why, in particular, is the epicenter of the Lamar Valley so perfect? For one, zoologists describe it as "an American Serengeti" because of the diversity and abundance of ungulates that gather there. Elk, moose, bison, mule deer, bighorn sheep, and pronghorn range across slopes that are richly endowed with edible grasses. In the winter when these animals die, they yield a vital source of protein that propels coyotes through the February breeding season and into the spring whelp (birthing). Where there is a profusion of grass, there is also a ready supply of rodents—voles, pocket gophers, mice and squirrels—that comprise the bulk of a coyote's diet.

Yet Lamar Valley is idyllic for still another reason equally as compelling as its natural smorgasbord: its topography. Essentially, the Lamar River drainage is a massive geologic cul de sac dozens of miles long and a third as wide. High-ridged mountains flank the perimeter of this ancient glacial trench. The natural architecture of the drainage discourages movement in and out of the valley by nomadic interlopers. Not to mention, the land is a buffer against humans outside the park who view coyotes as vermin.

Since the 1980s, Crabtree has built a comprehensive profile of the wild coyote following four years of research at Hanford, and a study along California's Mono Lake. He assumed correctly that the coyote territories of Lamar would be stable and easy to map. With forested uplands, river bottoms, semi-alpine meadows and arid benchlands, the mosaic represented the same diversity of habitats that coyotes had colonized in other parts of the country. From his base of intuition, he had not only predicted the correct number of coyote packs in Lamar, but remarkably their locations as well.

Although his study did not begin until just before the breeding season of 1990, Crabtree left the laboratory where he did his post-doctoral work at the University of California-Berkeley and journeyed to Yellowstone the first week of July 1989. Auspiciously, on a hillside located at the entrance to the expansive dell were an alpha male and female moving their pups from one den to another. Later in the year, Crabtree caught and radio-collared a yearling female that was part of the same brood. Before the study was completed, however, she would leave the Phantom Lake pack and try to establish her own clan. Such nomadic urges are called "drifting" and apply to individual coyotes that do not stay with their native pack.

Her sojourn took her forty-five miles due west as the crow flies to the shores of Hebgen Lake beyond Yellowstone's protective borders. The decision turned out to be a lethal one. She was found shot months later, which for Crabtree only strengthened Lamar's value as an ecological novelty set apart from a coyote-hostile world.

Often overlooked in the public view of coyotes is the animal's stunning beauty and indeed its close resemblance to not only wolves but human's best friend, the domestic dog.

Survival of the fittest. When there isn't enough food to go around, natural controls take over to prevent overpopulation. Disease often wields the final blow. The peak of pup mortality occurs between eight and fourteen weeks of age (mid-June through early August).

In many respects the society of coyote clans is nearly identical to wolves. The pack is held together by a single breeding pair—the alpha male and female—that are its nucleus. They exert primacy over their subordinates, the betas, by displaying dominant behavior which commands respect and obedience. Following the alphas in this hierarchical arrangement, betas perform a number of important duties, including helping to raise the pups, gathering food, scent-marking and defending the territory against usurpation by intruders. Next in line, at the bottom of the pecking order, are the older pups, or yearlings. On average, only two pups—out of the usual litter of between four and nine—survive their first year, but those that do become contributing members to the pack's cohesion.

While other animals might mate, the alpha pair is the only one in the pack that produces pups, though on rare occasions another female in the pack is allowed to breed. This is called a "double litter." Alpha males assert their leadership through a variety of forms and their faces bear the scars of tussling with rivals and enemies. They are the machine that drives the dynamic of the predator-prey relationship in the pack, whether it is undertaking the risky proposition of harassing an elk or bison in the snow or attempting to drive a grizzly off of carrion. He is also the burly protector of the turf. If a beta were to bark or woof to warn of a threat, out of nowhere the alpha male of the pack would appear in a swift charge. Stability in the pack is predicated upon his ability to find food and defend the clan. Being big and brawny and weighing almost fifty pounds, the alpha male of the Bison Peak pack was a fearsome warrior.

Crabtree noticed early on that the alpha male's partner, the Lamar matriarch, was assertive and individualistic in her own right. The anthropomorphic trait which makes canid alphas endearing to people is the fact that they routinely stay bonded for life. Such monogamy was present in the Bison Peak pack which established an impressive legacy. It had the biggest territory (four square miles), the largest average pack size (nine adults), controlled access to a long southwest-facing slope that yielded dozens of elk and bison carcasses in the spring, and portions of the finest grasslands adjoining the Lamar River that were dense with voles.

The matriarch and her companion never strayed far away from one another. "It is impossible to describe their interaction without drawing at least a few analogies to humans," Crabtree told me of the first winter when he saw her mousing in the flats on a cold January afternoon. She and her mate were out double-scent marking the fringes of the territory. She would urinate followed by the male who lifted his leg to mark in the same spot. The territory was theirs together. "They traveled in tandem eighty to ninety percent of the time," he added. "They would curl up on afternoons and sleep together in the sun, playfully taunt each other over a carcass and wag their tails in displays of affection."

THE PURSUIT

Behind us the ballooning inclines of Lamar Valley curve toward the foot of Bison Peak. Crabtree is in familiar terrain. Stuffed in his backpack are tools of a coyote researcher's trade: a good pair of binoculars, a thermos of hot tea, a change of raingear, sunscreen, a notepad to record field observations, plastic bags to inspect and collect scats, half a dozen radio collars and an equal number of Woodstream leghold traps. Long days are the norm for a biologist who trails animals active both day and night. Plenty of evenings he has remained in his plywood tracking station as the rising full moon summons a chorus of yips greeting the night. He would not depart the shelter until the gleam of dawn signaled the end of their foraging.

For an extensive study of coyotes to accomplish its objectives, it requires charting the movements of the animals over the entire span of numerous territorial ranges. A pair of naked eyes is good, the confirming tick of radio telemetry is better. Traps are a necessity because there is no other way—short of flying expensive helicopters—to strategically capture members of a given pack. It is important to note that the steel wares toted by Crabtree are not the average legholds used by fur trappers. Custom-built No. 3 softcatches are designed carefully not to break leg bones or harm muscle tissue. The jaws are padded and have shock absorbers to soften the pull of the trapped coyotes. Checked several times a day, they are made to hold a coyote just long enough for Crabtree to gently fit the animal with a radio collar and then set it free.

Although injury to coyotes can be high, Crabtree has never seriously harmed an animal while making over 200 trap captures stemming from his study at Hanford, Mono Lake and Yellowstone. It is his personal pledge, but more than that, the integrity of his research depends upon his care. To kill or maim even one adult in a pack by mistake would be inexcusable and have a ripple effect across the entire Lamar coyote community. "Our primary objective is

A wary coyote circles a trap set in a bison wallow. "I believe that ninety percent of coyotes that are eventually captured knew of the trap location many days before their curiosity was piqued and they dared to poke around at their peril," says researcher Bob Crabtree. "We think we're outsmarting coyotes and we're not."

Elusive and shy of humans, the matriarch of Lamar Valley anchored the largest and most effective pack of coyotes in the northern Rockies. Her genetic legacy is spread across a dozen packs. This rare photo of her a couple of winters before her death shows her robust physique.

being sensitive to the welfare of the animal," he says. "Injuring a coyote could easily alter its behavior as well as the results of the study, and we couldn't afford to let that happen."

Months elapsed before Crabtree and his team of field researchers came close to radio collaring the alpha male and female of the Bison Peak pack. "The alpha male is the hardest animal to trap there is," he says respectfully. Although leghold traps were objects that these coyotes had never known because wildlife in the park are protected, Crabtree was bemused at the methods alphas had of notifying him they were on to his modus operandi. Traps would be found turned over, sprung, and defecated on—just one example of how the coyote has earned its reputation as a cunning trickster.

"Radio collaring at least one member of the alpha pair is always a goal," Crabtree relates. "While you can learn a great deal about a pack by tracking betas and yearlings, the movements of the alphas are the real indicators of what is going on. Often, it is nearly impossible to fool a coyote in the center of its territory because they have a memory of how everything is supposed to look and smell." A large majority of captured coyotes come from outside, or on the periphery of their territory. Here, they allow themselves to investigate and be curious.

At Hanford, previous to his arrival in Yellowstone, Crabtree had hewn his boyhood trapping skills to the point that he was better than the professionals he enlisted. "In trapping, I let down the scientific guard and resort fully to intuition. It's just a feel you get," he said. Nonetheless, he will never forget a message he received from an alpha male who stymied him and was never captured during several years of vain tries. Crabtree was flying over the study area in an aircraft several hundred feet above the ground conducting a census. When the alpha male saw the plane approach, it began lunging into the air, snapping its jaws and openly taunting the human interlopers.

By the time Crabtree and his researchers were able to focus their efforts on the alphas of the Bison Peak pack, the meadows of Yellowstone were blooming with wildflowers. On May 10, he figured he was getting close because the morning brought fresh snow. Crabtree knew that if he got out at first light he could find tracks. "We followed them for half a mile and found mud stains on top of the snow and it hit me like a bolt of lightning. The bitch had been in the den and there was a mud path heading across the white snow. We tracked them another mile and they went right into a hole."

Before this quest to collar one of the Bison Peak alphas, Crabtree had already caught and collared 50 of the 130 coyotes that would comprise his study. It was an unspoken rule that traps would never be set within the vicinity of an active den, so he made three sets along the delta of the Lamar River and waited. Early the next morning, the research team checked the traps and struck paydirt. Two of the three sets made in loamy soil on the edge of the territorial foraging area yielded not only the alpha female, but her yearling daughter. It made the entire research crew exuberant.

A tenet in science is to do all you can to not traumatize or change the behavior of the animal you are studying. Crabtree does not believe in knocking out animals in traps with drugs. Rather, he puts a serious expression on his face and exerts dominance over the animal by making himself appear as big and looming as possible. Usually, the animals lower their head to the ground and cower. A rule is to never make eye contact because it can incite an aggressive response. He then gently wraps his hands around the animal and holds the head between his knees.

The only time Crabtree has ever been bitten came during one of his dominance displays as he approached a coyote in a trap. It was also a member of the Bison Peak pack, the son of the alpha pair. He stared at the canine and it looked at him, emitting a piercing, frightening howl that unsettled him. The shield of Crabtree's bravado was lowered and the coyote knew that it had exposed a weakness. As he moved in to take the animal out of the trap, it rose and bit him in the hand, rendering a scar that is clearly visible on his palm. Although he wasn't present when the alpha female was finally caught, he had instructed his field researchers to be cautious, so the same mistake would not be repeated.

On this day in the annals of Yellowstone wildlife research, the alpha female of Bison Peak was entered in the log as Number 570, a healthy, robust female weighing more than thirty pounds. After Crabtree performed an aging test by examining one of her teeth, he was stunned. The female was eleven years old and still turning out a brood of pups. The average life expectancy of a coyote in the wild is four years old, and the oldest ever recorded was sixteen in Colorado. "She was in shockingly good condition and fit into the upper one percentile of genetically fit beasts," he said. "We knew that she represented something special and she has never disappointed us."

Little did Crabtree know but despite all of the precaution and care he had taken not to disrupt the Bison Peak pack, there were forces beyond his control. Number 570 was about to become a widow.

With the return of the wolf in Yellowstone will the coyote be relegated to less-preferred forested habitats in the decades ahead? Only time will tell. Having occupied the niche of wolves for over 60 years, coyotes showed that their social structure is as sophisticated as their larger canid brethren.

PORTRAIT OF *CANIS LATRANS*

WE HAVE A LIGHT IN OUR FENCELESS BACK YARD that is activated by motion. The switch can be tripped by any sort of animation once the sun goes down. It might be set off by the wing-flapping of a moth or bat, for instance, or a shower of excessively large hailstones; or the bend of lilac branches being pushed hard by the wind.

Late in the night not very long ago, I was working with my thoughts drifting into the darkness when the florescent beam flicked on. From out of the shadows came the glow of two bronze-yellow eyes.

In a neighborhood with more housecats per square block than many places on the planet, I expected to see a friendly feline sashaying up the side-walk. Instead, it was a gray-ghost of a dog—a coyote—that breezed into the yard and froze still for an instant when it was struck by the unexpected shine. The animal's coat was grizzled; its tail bent toward the ground. The animal was out cruising, probably hunting for city cats.

Molded by nature into the perfect combination of fox and wolf, the coyote's long muzzle and perky ears have enabled it to be a highly efficient mouser, while sharp canid teeth and exceptional speed strike fear in ungulates.

Startled but bemused, I stood still. The visitor perked its ears and raised its nose to the air, sniffing for an olfactory means of explaining the sudden illumination, then it turned around and moved off quickly into the blackened alleyway. Was this an apparition? I wondered.

The next morning, I was more than pleased to find the confirming evidence. Imprinted in the snow was a staggered line of tracks with the noticeable indentations of four toes, dotted by claw points, and a rectangular heel pad. I took out a rule and measured them, three inches long, two inches wide, about the size of a small dog's.

"What are you doing?" my neighbor across the alley inquired as he walked his poodle past and let it spritz a fence post to mark its own proud, canid territory.

"Coyote tracks. I'm pretty sure I saw a coyote in the yard last night."

"A what?" he replied

"A coyote," I said.

"We don't want any of them around here," he assured me. "If they come back again I'll blow 'em away."

I feigned a smile, albeit a contorted one. With my sympathies offered here to the cats (and poodles) who now must prowl their kingdom in a heightened state of caution, I privately relish the fact that my back yard seems a wilder place now that the moon dog is skulking about somewhere nearby. I get the feeling that most of my neighbors do, too. As word of the coyote sighting spread, I received telephone calls from friends up and down the street, most of whom had moved to our tiny city from densely populated urbanscapes where nature lingered half-heartedly in parks and bird feeders. Yet they enthusiastically told me about coyotes inhabiting the hills near their former homes in Los Angeles and the outskirts of Chicago; the suburban golf courses of Houston and the urban wood lots of Atlanta. They shared my enthusiasm that the animals were here amid our little village of rectangular lawns as well.

Soon, I spoke with a handful of wildlife officials who had spotted coyotes in the fringes of dusk and dawn. Coyotes were blamed for the disappearance of laundry from clothes lines, missing house pets and the emptying of cat and dog dishes. The medicine wolf had come to life. By chance the next week, newspapers across the country splashed stories on their front pages about a pair of coyotes roaming the Bronx and dodging taxi cabs. It was a reminder

Ever resourceful, a coyote grabs a free meal off the front porch of a house. From Los Angeles to the Bronx and many places in between, coyotes are showing little timidity around civilization.

At the leap's apex a coyote has accurately located its next kill. Forelimbs, coiled like a snake, are ready to snap forward for the killing blow, to be quickly followed by a bite if the blow is not successful.

that this race of wild dogs are not fleeing from humanity but joining us on our own turf. "Coyotes are a lot closer than you might think," a game warden with the Montana Department of Fish, Wildlife and Parks once told me. "They're showing up everywhere, proving that they can live better beside us than we can survive next to them. It begs the question of who is more resourceful, humans or the cousins of the mongrels laying down at our feet?"

I'd go with the coyote.

Few people over the course of their lives will ever see a wolf in the wild, but the chances are good they will encounter a coyote. The dispersal of coyotes across North America has been largely a phenomenon of the last 100 years. Until the beginning of this century, coyotes were found in about half of the United States and primarily west of the Mississippi River; today they inhabit forty-nine states—all but Hawaii—but don't put it past them. You can find coyotes as far south as Costa Rica and northward to the Arctic Circle, from Mexico's arid Baja Peninsula and in the lush ancient forests of the Pacific Northwest to the chilly headlands of Nova Scotia and down the Atlantic seaboard into subtropical Florida. From Hudson Bay to the Rio Grande. They have ventured to these distant points and everywhere between. It is a remarkable tale of range expansion and we are witnessing it in our lifetimes. While many predator populations in the United States are in trouble, coyotes are thriving. Across North America, like in Yellowstone National Park, coyotes have probably tripled their numbers with the extirpation of the gray wolf and other predators.

Part of the coyote's ability to proliferate certainly is linked to its sheer intelligence, but nature has also given these omnivores the tools to beat back the best human attempts toward their eradication. Their body size allows them to function as scavengers and, if need be, to hunt in packs as full-fledged predators preying upon animals many times their own weight. The Coyote is an animal that has been given a bum rap, never quite measuring up in the public's eye to the glamour of the wolf. Routinely, instead of recognizing the beauty of Old Man Coyote we chastise him as the hapless villain who perpetually ends up with anvils on his head in Saturday morning cartoons. But the truth of the matter is, just as he had done for thousands of years with native people on this continent, the coyote is telling us something but we're not listening.

While the resemblance to their large canid cousins has earned them the title "little wolf," coyotes are superior to wolves in many respects because they are multiplying and prospering in places where wolves cannot. The very pressures of human population that have made wolves an endangered species south of Canada have made coyotes bountiful. It has spawned the "Super Coyote," a creature that has become ever more successful in evading humans yet finds itself living in close proximity to them. Between wolves and coyotes, the social structures and methods of communicating are parallel, and they share the same limb of evolutionary development.

There is a great deal of speculation about what came first, the coyote or the wolf. The fossil record, so far, has been inconclusive, though researchers believe that sophisticated samplings of DNA could yield answers. Going back through the taxonomic record, it is known that coyotes and wolves already were distinct species by the end of the Pleistocene epoch, roughly ten-thousand years ago. But before that, things are fuzzy. Maybe two million years ago, a mere tick in the time of the universe, a fork in the road of canid evolution appears to have given them separate identities.

Until that point, some taxonomists believe that a larger ancestor coyote emerged from the animal that is the ancestor to today's foxes. In turn, it has been proposed that wolves then split off from coyotes. The conventional and more accepted theory is that wolves came first, but the bare-boned facts have yet to prove it conclusively.

Archaeologists have now unearthed coyote bones at digs across the country that are older than the earliest dated wolf remains. Wolf researchers believe the coyote is merely a primitive offshoot and a species of far less social refinement. Some of those who study coyotes say that *Canis latrans* is just a creature of different, not necessarily lower, development. But just for fun, think of the possibilities. If the fossil record is proved to be accurate by DNA—and that is a big, big if—we humans accustomed to judging the value of things by their physical size may have to approach these canids from a whole different reference point. Instead of calling coyotes "the little wolf," we may be tabbing wolves as "the big coyote." Nonetheless, speculating about the coyote's evolutionary origins is a fascinating exercise because it helps to reveal where the coyote sits, morphologically speaking, in comparison to other animals within the order of Carnivora.

Surprisingly similar in size, a coyote skull is displayed in the foreground with a wolf skull behind it. Archaeologists so far have uncovered fossils from coyotes that are older than wolves.

Many people often mistake coyotes for wolves and indeed the two species share remarkable behavioral similarities. Wolves are up to three times greater than coyotes in weight but are scarcely half-again as large in frame dimensions. This wolf is a healthy example.

Carnivores need special physical tools to kill and devour their prey. Most canids have forty-two teeth and the trademark front canine teeth. These fangs allow the animals to puncture flesh, while premolars are used to tear off the meat, and back molars serve the purpose of crunching bone.

One of the subgroupings under the umbrella of meat-eating animals (that includes bears, cats, and weasels) is the family Canidae, which represents dogs. Under Canidae with the genus *Canis*, you find the breeds we see everyday barking at the mailman and retrieving our ducks, *Canis familiaris*, the domestic dog. This same general family is the one from which the coyote diverges to claim its novel place among wild dogs in the world. *C. latrans* is closely related to the domestic dog, as well as the wolf (*Canis lupus*) and also the Australian dingo (*Canis familiaris dingo*), the red wolf (*Canis rufus*), three different species of jackals and more distant ancestors, the foxes—the common red (*Vulpes vulpes*), gray (*Urocyon cinereoargenteus*), and rare swift (*Vulpes velox*).

True dogs of the Americas, coyotes are bona fide natives dating back to time immemorial. Wolves, on the other hand, appear to have traveled back and forth across the ancient land bridge between Alaska and Eurasia known as Beringia, which explains the reason they are found in boreal and arctic regions across the globe. The coyote, meanwhile, evolved in North America and here it stayed, never leaving. Although endemic to North America, it is physically and closely linked ecologically to the golden jackal (*Canis aureus*) of Africa, and in the southern United States has been mistaken for the red wolf, whose range in recent historic times extended from Texas into the Appalachian Mountains.

During the last decade, the U.S. Fish and Wildlife Service has undertaken efforts to try to restore the red wolf into its former niche, but a great deal of debate exists about whether the survivors are really full-blooded wolves or actually coyote-wolf hybrids. " . . . it seems reasonable to suggest that when enough evidence is gathered the red wolf will be found to be properly known as *Canis lupus* X *Canis latrans*," writes renowned wolf biologist L. David Mech in his classic work *The Wolf*. "In other words, the red wolf may be a fertile wolf-coyote cross, with eastern races of the hybrid population more strongly resembling the wolves that once occupied the forested Southeast and with whom they periodically bred; and with the western race of the hybrid populations more strongly resembling the coyotes that still occupy the open area of Texas and with whom they still may breed."

The question has implications for coyotes and the much-larger gray wolf.

Interbreeding is a problem where the numbers of one species fall into such decline that they have no other choice but to pursue their strong instinctive breeding with close relatives. Earlier in the twentieth century when red wolf populations were severely reduced, biologists believe that lack of available breeding animals and the fact that the preferred habitat of red wolves and coyotes intertwines probably resulted in inter-species breeding. The same suspicions apply to the larger coyotes ("brush wolves") of Minnesota, southern Canada, and the Northeast.

"BARKING DOG"

Regardless of where one encounters coyotes on the continent, the animals have universally identifiable features and markings. The pelage of their coat is usually a silver or grizzled gray, particularly during the winter months, but the fur can change subtly in the summer. The legs, ears, and snout tend to be tinged with rufous and ocher hues. In terms of their size, it depends upon the region of the continent they inhabit. Geography shapes the size of the animals as well as the size of the prey they have evolved to hunt.

In all there are at least nineteen subspecies of *C. latrans*, the largest being the eastern woodland coyote found in northern New England, where individual animals have surpassed sixty-five pounds, making them comparable to a small wolf and of questionable genetic purity.

The speculation is that perhaps the size of eastern coyotes may have resulted from inbreeding between coyotes and wolves. The phenomenon is not limited to wild canids. There also is a hybrid that occurs when coyotes crossbreed with domestic dogs. This "coydog" can take on a strange appearance given the breed of the animal the coyote chooses as a mate. When coyotes were killed recently in the Bronx, wildlife officials discovered a coydog roaming a nearby cemetery, confirming that crossbreeding is taking place. Nonetheless, the sightings in the Bronx again propelled the coyote into near mythical status.

Native peoples in the West had known of the magical charm of coyotes for perhaps 10,000 years or more because there are petroglyphs and pictographs that testify to the reverence held toward them by ancient artists. The Old World first came in contact with coyotes when Spanish conquistadors noticed them in Mexico and the Southwest as early as the 16th century. In 1651, Francisco

Still pups but three-quarters adult size in September, they practice howling. Pups are now joining adults in territorial vocalizations and can be distinguished by their high-pitched tones.

Hernandez gave a somewhat accurate but deeply embellished portrayal of the beast in a natural history citation. Hernandez described the coyote as "an animal unknown to the Old World, with a wolf-like head, lively large pale eyes, small sharp ears, a long, dark and not very thick muzzle, sinewy legs with thick crooked nails, and a very thick tail. Its bite is harmful." Other written accounts of this uniquely New World dog appeared with Meriwether Lewis and William Clark when they came up the Missouri River, though it would take two decades after their adventure before the coyote made its official arrival in scientific nomenclature.

In 1823, naturalist Thomas Say took credit for naming coyotes after he encountered them near the current site of Council Bluffs, Iowa a few years earlier. His observation came while in the company of Major Stephen H. Long who was on an expedition to map the frontier. Say was struck by the animal's smart ways as well as its frequent and varying vocalizations which were distinct from the howls of wolves. As a result, he dubbed coyotes *Canis latrans*, Latin for "the barking dogs."

Coyotes in the West, with the exception of the Yellowstone area and north to the Canadian province of Alberta, tend to be much smaller and more representative of the typical beast. Twenty to thirty-five pounds is a common weight. During the nineteenth century, Lewis and Clark likely saw glimpses of what some people colloquially call the "prairie wolf" (*Canis latrans latrans*) and this is probably the species that Say recorded as the barking dog. *C. latrans latrans* is another small coyote subspecies.

Larger in frame and weight are the coyotes of Bob Crabtree's study in Yellowstone (*Canis latrans lestes*) which are the romanticized "moon dogs" synonymous with the Rockies and desert region of Colorado, Utah, Arizona, and New Mexico. One hypothesis pertaining to the physical stature of Yellowstone's coyotes is that over the past half century each new generation has been slowly growing in size in response to the absence of wolves, and the need to feed on large game species. The evidence is anything but conclusive, and regardless, the continued growth spurts are certain to be short-lived. Wolves are returning to Yellowstone, casting coyotes back into a secondary role.

In other regions of the country where wolves and coyotes have shared space, it has always been coyotes that fall back into a secondary role, if not to disappear altogether. Durward Allen, the international expert who wrote the definitive book about Isle Royale's wolves, says it is not a question of how coyotes will be

In addition to the coyote's excellent hearing, its acute vision aids in the detection of minute movements of prey through grass.

displaced but when. Wolves in Yellowstone are certain to rearrange the canid pecking order either by thinning out coyotes or extirpating them. At Isle Royale, located off of Michigan's upper peninsula, coyotes had thrived on a steady diet of small rodents and moose carrion without any competition from a large carnivore. "Wolves got to Isle Royal probably in 1949 during the winter time by crossing out on the ice from the mainland," Allen, now an octogenarian, told me by phone one evening from his home in West Lafayette, Indiana near Purdue University where he taught for many years and pioneered the study of timber wolves at Isle Royale. "After they got there, it was five or maybe six years until someone saw the last coyote track. It appeared that the wolves just plain killed them off. There is competition among carnivores near the top of the food chain. If coyotes in Yellowstone are displaced by wolves, it would not surprise me. That's what most of us think will happen."

TERRITORIALITY

According to Crabtree, many of the unexploited coyote packs he's observed in the West have exhibited wolf-like approaches to hunting and territoriality, which promises to make things rather interesting when coyotes and wolves eventually vie for territorial control again in Yellowstone. How it will shake out, in terms of how coyotes reorganize their territories, is anybody's guess. What exactly is a territory?

"The territory of an animal is defined as the areas that the animal will defend against individuals of the same species," writes L. David Mech, the noted wolf biologist in Minnesota. "The defense of the area is the main difference between a territory and a range or home range."

Coyote territories are stretched out invisibly toward the horizon line contiguously and in overlapping fashion. If they could be viewed from the sky they would appear hexagonal, with six territories woven around a central one, and so on and so forth. Hexagon connected to hexagon, over and again. Wherever there is a group of coyotes there is likely to be a pack cell, and thus, a territory. In between, working the seams are small bands of nomads passing in and out that may have no genetic connection to those in the packs. It is these animals that we generally think of as the stererotypical lone coyote. The amazing aspect of territories is their longevity and apparent permanence. Crabtree does not believe they

A coyote surveys his domain in the exposed earthen badlands of the American West. There isn't a state in the contiguous U.S. that these adaptive canids have not colonized.

are set up randomly but that the landscape dictates their dimensions.

Crabtree says territories can transcend the life of individual packs. Think of a territory as a series of concentric circles, or better yet, a target with the highest number of points in the center and diminishing values the farther out you go. That's the way it is with coyote territories. The middle of the territory often envelopes the place where coyotes position their natal dens and will be fiercely defended, the interim zone is prime foraging ground that usually rates only growls between locals and invaders, and on the frontier of the territory the perimeter usually overlaps a coyote territory from the other direction.

A question tugging at researchers is whether coyotes are losing their ability to form packs as they lose the wild open spaces that support them. "However characteristic pack-units may have been of *Canis lupus* under aboriginal conditions," wrote J. Frank Dobie is his classic *The Voice of the Coyote*, "the animal became increasingly solitary after man broke into 'Nature's social union.'" Human intolerance and erosion of the wilderness certainly didn't help.

VOCALIZATIONS

"The voice of the coyote is one of its most remarkable gifts," wrote Ernest Thompson Seton in *Life-histories of Northern Animals* nearly a century ago. "Barking is supposed to be limited to the dog and coyote. This is not strictly true, for wolves, foxes, and jackals bark at times, but it is true that the coyote is the only wild animal that habitually barks . . . Most of the many calls of the coyote are signals to its companions, but some of them seem to be the outcome of the pleasure it finds in making a noise. The most peculiar of its noises is the evening song, uttered soon after sunset, close to camp."

Coyotes are a lot like wolves in their reliance upon vocalization as the chief mode of communicating to others of their kind. Most people hear coyotes before they see them, but every bark, yelp and yip is a lexicon in itself. Few choruses are as haunting and yet sweetly tranquil. At the National Elk Refuge in Jackson Hole, Wyoming, Franz Camenzind spent several years listening to the language of coyotes and correlating it to behavior. "Communication is essential to maintaining order within socially organized populations of animals and the most conspicuous form of communication employed by coyotes is their vocalizations," he wrote in his doctoral thesis "Behavioral Ecology of Coyotes."

The lone howl is a vocalization that coyotes often use to locate other members of the pack. It is heard routinely during the winter months across the territory.

A dominant alpha male (with straightened tail) approaches a trespassing intruder (with tail in tucked or submissive position). The alpha female and ten-month old young of the year serve as backup. Alpha males are typically initiators of all confrontations whether with killing prey or territorial disputes. The message was sent and the resident pack wandered off, unlike wolves which often kill each other.

Camenzind identified and described nine vocalizations. These were divided into three categories based upon their functional and phonetic relationship. The first category included the growl and yelp and were produced during severe agonistic encounters. The second category consisted of the woof, bark and bark-yip that formed a functional (warning-challenge) relationship as well as a phonetic sound continuum. The final category consisted of the lone howl, greeting song, group howl and the group howl-yip. The first two communicated information exclusively within packs while the last two functioned both with and between adjacent packs. Along with Camenzind's work, biologist Philip Lehner has identified two additional vocalizations, and as the animals continue to be observed there are bound to be more.

The primary sounds delineated by Camenzind that may be familiar to listeners are the following: The *growl* is a gutteral sound seldom repeated more than two or three times and is often produced by an aggressor during antagonistic encounters. It is an expression of dominance. The *yelp* is a sharp high-pitched "i-eee" sound repeated two to six times and slightly louder than the growl, which conversely is an expression of submission. The *woof* is a muffled woof or barking sound emitted by the adults in the vicinity of pups, perhaps as a warning call to find a place of safety.

The *bark* is another adult-only expression, that resembles the bark of a medium-sized domestic dog in sharp, quick intervals. It is a vocalization apparently signaling alarm and challenge and might be expressed for up to 30 minutes at a stretch. The *bark-yip* is an adult warning call that Camenzind heard while walking his dog through a coyote territory. It was accompanied, he said, by intense aggressive body postures from a coyote with its tail held from horizontal to 45 degrees. Eventually the intensity subsided somewhat and the offended coyote sat and continued barking and yipping.

The classic *lone howl* is a long drawn-out call produced by a single coyote once or several times in succession and is the lowest pitched of the three types of coyote howling. It is mostly heard during the winter months and, to a lesser extent, in the summer usually to locate separated pack members. Pups, too, might perform this howl as an emulation of their parents.

The *greeting song* is a modulating soft howl that is audible up to approximately 900 yards and emitted almost as a contented statement when pack members were reunited. It is performed as a part of the total greeting ceremony. The *group howl* is a series of calls made exclusively by pack coyotes which last from 30 seconds to two minutes. The sessions commence with a deep, drawn-out lone howl with other pack members joining in what may be a chorus response to adjacent packs informing them that the territory is occupied.

The *growl yip-howl* is a similar variation to the group howl, but involves an overall increase in excitement that includes high-pitched, sharp yipping at the height of the session any may involve pups. This vocalization—more than any other—elicited similar responses from adjacent packs in Camenzind's studies. Consequently, he frequently located packs by imitating the frenzied howl-yipping, particularly by the defenders in the case of actual boundary crossings. All such encounters included high-pitched yipping, tail movements, and stiff-legged, abrupt body movements.

Without exception it is a treat to put oneself in a position to hear the beautiful vocal interaction between and within coyote groups. But until a human becomes fluent in the coyote language and is able to pinpoint the meaning of specific nuances, much of the deciphering beyond the general correlation between sounds and documented behavior will be speculation. There is always more to be learned. For the casual human listener, however, it only entices to want to come back for more, particularly when the songs of the coyote blend in to an evening or early morning symphony.

SCENT MARKING

When humans want to keep people out of their property, they erect a "No Trespassing" sign. For coyotes, the task is accomplished by scent marking, which involves depositing scat and urine throughout its territory. Every pile of coyote droppings and urine carries the personal signature of the individual animal. Coyote dung is marked as it leaves the animal's sphincter by rubbing against a pair of anal musk glands. This provides information to other members of the pack and nomads drifting in and out of the area.

Coyotes, like all dogs, seem almost compulsive about depositing their own sign and studies have shown that next to hunting, as much time in the average day is spent scouting the territory and re-scenting than anything else.

Positioning of scat and urine occurs strategically in the wild, too, often along game and hiking trails, and especially where the paths crest a hill. Logs are attractive to coyotes as "scent posts" for depositing urine but the animals often choose

A coyote checks the scent on a tree marked previously by a male. Urination occurs at a much higher rate on the periphery and boundary of the territory.

Coyotes can adapt to a wide variety of environments, from cactus fruits in the deserts of Mexico to arctic ground squirrels in the tundra of Alaska.

prominent rocks, trees, and even glades of grass. Over the course of a day, the same place might be marked by many different coyotes. The action is especially frenetic on the border of two adjacent coyote territories.

There is, believe it or not, a highly refined approach that a coyote takes to this. It is not done willy-nilly. An animal will arrive at a site, let its nose inspect for other markings, then either lift its leg or squat depending on the sex of the animal.

"Coyote scats are extremely variable in size," wrote Olaus Murie in *Animal Tracks*. "The residue from pure meat is likely to be semi-liquid. The scats consisting of much hair are likely to be large. Those resulting from a diet of pine nuts or chokecherries are likely to crumble. In size, coyote scats overlap those of the wolf and the red fox, and those of pups are of course smaller."

The shape, texture, and composition of coyote droppings change according to what the animal eats. When solid in form, the scat is about the size of a small dog's, perhaps four inches long and with a curl-twist at the end. After voiding dung, the animal rakes at the ground, which sometimes results in distinct scratch marks—this behavior is also exhibited by domesticated breeds of dogs.

Scent marking is certainly a means for other coyotes to decipher information, but precisely what that information is has yet to be fully realized. Crabtree has several theories. It might indicate whether it is a good food year in a given territory, how many individuals might be present in a given pack, and the social hierarchy. During an innovative study at the Arid Lands Ecology Reserve in Washington State, Crabtree implanted coyotes with a chip that released a trace amount of radio isotope into his subject's scat and used it as a means of defining how coyotes employ scat to delineate their range. He found that clumps of scat distributed around prime foraging areas send a signal to other coyotes that there is no food there, and that it may send a signal to outsiders that some other coyote had already staked a claim.

DIET: THE GENERALIST CONSUMER

Patterns emerge in nature to minds that are willing to consider grand designs with higher purposes. In snowflakes and tree rings and fish scales, we find remarkable symmetry. When evolution drew up the blueprint for the canids of North American, it gave wild dogs an uncanny kindredness through proportion.

Red foxes are roughly one-third the size of coyotes, and coyotes roughly one-third the size of wolves. With foxes, the long, sure-smelling snout resulted in the consummate mouser. With wolves, the large bodies, powerful jaws and swift speed are potent accessories for bringing down big herbivores. Coyotes fill the niche between both species and possess what has been called "biological reach." Depending upon the sources of prey that avail themselves, coyotes can subsist solely on rodents and insects, or they can band together as hunters of an animal fifteen times their own weight. Most humans don't recognize that coyotes are among the fastest land animals on the continent, capable of reaching speeds of forty-five miles per hour over short bursts. It explains the animal's skill in catching pronghorn fawns on the open plains.

In his seminal study, "Ecology of the Coyote in Yellowstone," Adolph Murie assiduously collected 5,086 coyote droppings and analyzed the contents. Within the excrement, he counted 8,969 individual food items. More than 3,000 of the contents, or almost thirty-four percent, were the remains of field mice, twenty-one percent were pocket gophers, and over sixteen percent were adult and juvenile elk (consumed mostly in the late winter and spring). In northern wildland settings during the winter months when rodents become scarce, they turn to ungulates, otherwise they catch dozens of mice and gophers each day. These are the staples of the Yellowstone coyote, but intermixed in the scat and depending on the season, Murie found the remains of grasshoppers (which accounted for seven percent of his subjects' diet), nearly three dozen other large and small mammals, a dozen bird species, a fish and garter snake, and an assortment of vegetative matter ranging from pine nuts to blueberries.

Coyotes are ingenious in exploiting whatever is available. This makes them "generalists." In the Southwest, in addition to mice and hares, they feast on dozens of things from lizards and roadrunners to the soft vegetative meat beneath prickly pear cactus; in the Pacific Northwest, they may eat fallen fruits in apple orchards; in Los Angeles, rats, chihuahuas and housecats; in the Midwest, crawfish and porcupines; in the Southeast, they have raided patches of watermelon; outside of New York City, the contents of sanitary landfills. Researchers in Alberta noted that elk and, to a lesser extent, deer and other ungulates accounted for sixty-seven percent of a coyote's food.

Murie's findings are revealing because they point out an interesting array of scat contents beyond what we might expect from a somewhat pristine

Playing with its mouse prey, this coyote tosses its next meal airborne only to catch it again like a toy.

environment like Yellowstone. Here are just a few of the miscellaneous and non-food items that passed through different coyotes' gullets: horse manure, muskmelon, a canvas leather glove, twine, banana peel, shoestring, eight inches of rope, fragments of towels and shirts, tinfoil and cellophane.

The coyote will consider eating whatever is put in front of its mouth. "The coyote's favorite food is anything he can chew; it does not have to be digestible," wrote Frank Dobie. In lean times it has learned to not be finicky. This is why the coyote can exist as easily in a wilderness like Yellowstone as the back streets of New York City.

Attacks on humans by wild coyotes are extremely rare. However, when they become habituated to people who feed them in national parks, they can turn aggressive, and commonly inflict injury.

CHAPTER 3

THE SOCIAL ORDER

ONE OF THE GREAT PUZZLES OF CANID RESEARCH over the years revolves around the question: In a wild environment, are coyotes naturally predisposed to group in packs, or are they more solitary—as we hear about in the old cowboy legends? If the former is true, it means that wolves are not the only sophisticated social predators on the continent and coyotes have been underestimated. If the latter is correct, then maybe the modern stereotype of the coyote as a lone and solitary beast is an accurate portrayal.

As Bob Crabtree stood in the summer twilight of Lamar Valley and perked his ears to the dueling yip-howls of rival coyote packs around him, a grin crossed his face. He stretched out supine in the grass with his arms splayed behind his head and closed his eyes.

Crabtree imagined a scene from the nineteenth century as recorded by explorers in their journals about hundreds of coyotes being visible on the frontier in the course of a single day. His thoughts drifted next into the realm of the

The singing trickster, long a fixture in the mythology of Native American tribes, is valued for its cleverness and ability to turn difficult situations to its advantage.

so-called medicine wolves (a Native American name for coyotes) that trailed horseback-riding warriors and feasted on gut piles of buffalo left in the wake of successful hunts.

There had to be reasons, Crabtree thought, that would make people routinely mistake coyotes for wolves. The Bison Peak pack soon would make it all seem clear to him.

PREY

As he settled into relaxation, Crabtree's thoughts were interrupted by the crackling ticks of coyote No. 570—the alpha female—re-entering the range of his field receiver. She had been wearing a radio collar for only three months following her capture but already and unknowingly she was offering Crabtree a vicarious invitation to fall in with a great chase. He hopped up, letting the radio ticks be his guide.

Now that her pups were twelve weeks old, the canid matron had turned more mobile again. With her alpha mate guarding the pups, she and one of her daughters, a beta, had spent previous evenings meandering through the buffalo grass along the river, poking their snouts into the runways of voles to drum up supper.

Over the trunk of a fallen cottonwood just the night earlier, she had let out a bark and wagged her tail, entreating a wary chipmunk to test its luck. Wisely, the rodent remained hidden. This time, No. 570 decided not to press the matter. The extra clump of meat and fur wasn't needed because her belly was already full.

Every evening at dusk, the bitch faithfully brought home the spoils and regurgitated them into the mouths of the seven hungry pups in her new brood. Tonight, however, the scene was different. Although the other six adult members of the Bison Peak pack were on territorial patrol, the alpha male and female had assembled two available betas, leading them on a foray into the higher mountain meadows.

The light was dimming enough so that Crabtree knew he had little time to waste. He set out at a trot. Shadowing the pack at sundown came with certain risks for a researcher. There were grizzlies on the slopes, overprotective sows with cubs grubbing for roots and most likely pursuing the same type of forage as the coyotes.

The usual alarm call from this prairie dog colony in North Dakota did little to protect one of their own from the wily coyote.

Quarrels over carcasses, particularly during lean winter months, are common within coyote societies. But seldom do such snarling matches over food ever produce serious injury.

Long before Crabtree's study commenced in Yellowstone, the Bison Peak pack had gained a reputation among park rangers. They knew that this band, among the other twelve spread out across the valley, was special. These smallish wild dogs, you see, were elk killers, accomplishing what some researchers had attributed only to wolves, bears, and mountain lions. A few winters past, park ranger Joe Fowler had discovered a mature, 500-pound cow elk that was partially eaten with lethal marks that indicated the handiwork of coyotes. He could reconstruct the story of the chase and the details of the attack in snow: rear legs were punctured by teeth that tore the tendons, started a little cut in a big vein and ended with a bite to the jugular. How could it be that a forty-pound animal had conquered a giant nearly thirteen times its own weight?

"The survival of an animal depends critically upon its ability to find food and to exploit it adequately," wrote William Donald Bowen in his thesis called "Social Organization of the Coyote in Relation to Prey Size." "It follows then that the social organization of a large mammalian predator with few natural enemies may be primarily an adaptation to its food." Crabtree found larger packs in the Lamar Valley because there was an abundance of large prey. In the more than sixty years that coyotes had thrived in Yellowstone without wolves around to bother them, they had solidified their social structures and developed the skills necessary to meet giant ungulates on their own terms. In other parts of the country, coyotes had indeed been known for their ability to prey upon deer and antelope but many canid biologists assumed that taking on elk and bison was unlikely.

Crabtree had no preconceived notions when he began his study in Yellowstone. He wanted simply to let the packs themselves deliver an answer.

And it didn't take long for him to discern that, if given an opportunity, coyotes can and do fill what traditionally had been the niche of wolves. This raises an interesting comparison about which species is most adaptable and here again, the coyote comes out on top. In *Never Cry Wolf*, writer Farley Mowat portrays wolves as being able to get by without large ungulates. In Mowat's work, wolves eat mice. However, most canid biologists say it would be virtually impossible for a canid as big as the wolf, which demands at least several pounds of meat a day, to sustain itself merely by consuming small rodents. With coyotes, it is a different matter because an individual can get by on a third as much food. So it goes without saying that if there are not sustainable numbers of ungulates available, wolves cannot persist. In other words, they cannot become coyote-like.

On the other hand, as Crabtree observed, coyotes can rise to the challenge of becoming wolf-like in hunting large prey. Though they certainly are not nearly as efficient, larger coyotes, like those being produced year after year in the Bison Peak and surrounding packs, can fine-tune their predatory behavior to take down elk at the times of year when it is most necessary—the long winter months when coyotes are shut out from voles and mice by a layer of ice on the ground, and summer when pups need to be fed.

When it comes to stalking large prey species, whose sharp hooves represent a lethal defensive weapon, coyotes are brilliant strategists that know how to play the waiting game. Former Yellowstone park ranger Bob Murphy told me that he witnessed a case where a group of coyotes years ago ran a young elk into a river but did not follow. They waited for seven days. When the fatigued cow elk finally walked out of the current, the coyotes brought the animal down and consumed it. Bert Harting, a grizzly bear biologist, has no doubt that such confrontations occur. During the 1980s, he conducted an innovative study where elk calves were radio collared and followed to see how many were preyed upon by bears. Seventeen percent, he discovered, fell prey to coyotes.

Crabtree suggests that 1,000 elk, probably more, are killed by coyotes every year in Yellowstone, far surpassing the number taken individually by grizzlies and mountain lions, making coyotes the most prolific killer of large ungulates in the park. More profoundly, coyotes have been making inroads on bison, with Yellowstone park ranger Les Inafuku having documented the first example of a coyote-killed bison in North America. Crabtree and his researchers have observed several unsuccessful predation attempts on bison in Lamar Valley. In one instance, an alpha male attacked the hind quarters of a 1,000-pound bison.

The more hours that researchers spent surveying conditions in the vicinity of Bison Peak, the more that the remarkable legacy of the resident canids began to emerge. They knew that predation was happening but it is difficult to chronicle given the secretive nature of coyotes and elk. "Direct observation of coyote predation on large wild prey are rare," wrote Eric Gese and Scott Grothe in *American Midland Naturalist*. "We observed nine predation attempts on deer and elk." Gese, from the University of Wisconsin, and Grothe, from Montana State University, were both graduate students working under Crabtree. Grothe noted that the alpha male led attacks on ungulates nearly ninety percent of the time, demonstrating the important role of alpha

During the middle of a bitter winter, patience is a virtue. For the coyote, it knows it has meat on the hoof with this mortally wounded elk. The protein yielded by this kill, equivalent to thousands of mice, will help carry members of the pack and other scavengers through to the spring.

East of the Mississippi, coyotes have replaced wolves as common predators of the white-tailed deer.

These photographs capture a rarely-seen sequence of events documenting a single coyote as it pulls down an ailing buck.

males not only in helping to provide food for other members of the pack but in teaching behavior that is passed down to future pack leaders.

Several things can be deduced. Throughout North America where the average pack size is four or five coyotes, the animals have evolved like African lions where the pride size functions primarily to protect the young and thus perpetuate the group. Similarly in Yellowstone and on the National Elk Refuge, pack sizes are bigger because fewer pups disperse from packs the first year. The impetus is that having more adult and sub-adult animals means there will be more coyotes to identify and protect carrion within the territory.

Since the alpha female is one of the first animals to benefit energetically from having secured food sources at critical times of the year, that translates into a successful pregnancy and healthy pups. While in other parts of the country, pups often disperse from packs at six months, Yellowstone alpha males and females do not run them off before their first birthday and occasionally regurgitate to them at this late age. The alpha male and female obviously realize the benefits of having them stay in the pack. The young betas learn much about future parenthood and help defend their younger siblings. With coyotes, everything comes back to maintaining livable conditions for pup survival.

The reign of the Bison Peak pack arose as no accident. Its strength can be measured by the sum of its members, each of whom was taught by the alpha pair. This is precisely what is meant in nature as good breeding.

As Crabtree cleared the hill, he sensed commotion from the area of a small creek where the loud signals from No. 570 were emanating. Amid a stand of aspens, he saw a group of coyotes and a frantic band of cow elk. He would not approach closer because he didn't want to interrupt the interaction. The problem was that dusk had passed and it was too dark to use his field glasses.

"If you want to learn about coyotes, you have to live with them, sleep when they sleep and be active when they are active," Crabtree says. "You have to conform to their schedule." It sometimes means starting the research day just before sunset, being attentive through the night howling and hunting, tracking them past midnight, relaxing about 3 a.m. when coyotes take a siesta, readying for the morning hunt, and listening to songs that signal an end to their nocturnal ramblings at the first hint of sunrise.

He stayed out all night and the next morning wandered over to the scene of the confrontation between the elk and the pack. Picked almost shiny clean was the rib cage of a young calf that had been brought down by the alpha male. The meat would sustain the Bison Peak pack for the next few days and help ensure that the pups would survive.

The genesis of a new litter of pups begins in December with the arrival of colder weather pushed southward from Canada. For weeks prior to breeding, the alpha male and female spend nearly every minute together. Double scent-marking is an integral part of the courtship. Before estrus, the alpha female typically squat urinates on the territorial boundary, then after sniffing, the male lays down a squirt in the same place. They do this for hours, circling the perimeter of the territory. This is the quintessential indicator of an alpha pair in courtship behavior. (After she is impregnated, the male tends to urinate first and she follows.)

In late January, the bitch shows preliminary signs of going into heat. By the middle of February, the pair is ready. There is only a narrow window for pregnancy to occur as the female ovulates only once a year, and she repels advances until her body tells her she is ready. During copulation, the male mounts her from the rear. This "copulatory tie" lasts for fifteen to twenty minutes but it may be repeated several times.

The period of gestation in coyotes is relatively brief, only sixty-three days, and if mating is recorded, the whelp can be predicted within a matter of days. Every year, courtship, mating and the birth of pups occurred like clockwork within days of the previous season. Normal sizes of coyote groupings in the Lamar average seven adults and range from three to eleven individuals. Adult members of the resident packs are much larger than the twenty-five-pound average found among coyotes in the rest of the inner West.

DENNING

An important phase in a coyote year is denning and it is replete with its own steadfast rituals and customs. In 1945, Weldon Robinson was assigned by the Denver Wildlife Research Laboratory (an arm of the Fish and Wildlife Service) to complete a study on the movements of coyotes migrating in and out of Yellowstone National Park. Nearly half a century after he began his study, Robinson returned to Yellowstone at Crabtree's request. Crabtree had

Radio collared mates on the Blacktail Plateau in Yellowstone National Park engage in the classic canid "tie-in" which lasts on the average 15 minutes. It is believed that this long copulation sequence evolved to ensure impregnation.

an interesting query for the octogenarian. He asked Robinson to locate den sites that were previously identified in the 1940s following the pioneering field work by Adolph Murie. Some of them, both scientists concurred, may even have been abandoned by wolves which were essentially extirpated from the park in the late 1920s.

Not only could Robinson remember the exact position of old coyote dens, but much to the surprise of both he and Crabtree, they found that the subterranean chambers still were being used by alpha females to birth and nurse their pups. Crabtree determined that at least three-fourths of the den areas originally mapped by Robinson were active, which provides at least a sixty-year testament to the level of fidelity that coyotes hold for strategic parts of the landscape.

Much to their surprise, the first successful capture of pups that Weldon had pointed out in the spring of 1946, was the exact same den site Crabtree had just captured pups in some 46 years later. Maybe not so surprising was the fact that the den site belonged to the Bison Peak pack.

The matriarch was loyal to other den sites, too, and the placement of her pupping areas is typical for where breeding females choose to remain during the last few weeks of spring pregnancy. That spring, the Bison Peak pack had eleven adults heading into the pupping season and was supplemented by a double litter born to the alpha female and one of her beta associates. It elevated the local census to twenty-three. This would prove to be the last gift of genetic stock that both the alpha male and female would make, but that is getting ahead of the story. For now, mother, daughter, and granddaughter were together near their respective dens.

Den sites can undergo their own metamorphosis. They are not year-round domiciles but temporary maternity wards and nurseries. Provided the alpha male and female do not erect the den themselves, a den may start as nothing more than a ground squirrel hole tunneled into a south-facing slope surrounded by cheat and bunch grasses. Along comes a badger who smells the rodents inside and, with claws extended, broadens the mouth of the future den by excavating its prey. If the site is situated near other rodent populations, the badger may decide to start a new den. In either case, whether this underground passage is left vacant, or whether the badger takes up residence and abandons its old den site, a coyote mother often realizes the dividends by claiming one of the holes.

Birth in coyote packs occurs usually in late April or the first week of May. Typical coyote litter sizes can range from four to eight. The largest on record is 19, probably the result of a "communal litter."

In late June or early July amid the lushness of summer vegetation, pups seek out shade around dens or in nearby rendezvous sites.

rearing, Hatier's hours of field observations showed. They feed the babies through regurgitation, groom them, and in some cases the females even begin to lactate and assist the mother in feeding her young. It is really an apprenticeship in pup rearing and the coyote's way of preparing underlings for the future when they either set out on their own to form a new pack or ascend to alpha when the matriarch or her mate dies.

Mortality negates any gain in a pack and there are some years when few pups survive. Out of an average litter of six, only two pups survive the summer in unexploited coyote clans. When there isn't enough food to go around, natural controls take over to economize. Disease often wields the final blow. The peak of pup mortality occurs between eight and fourteen weeks of age from about mid-June through early August.

Playing with pups is essential to the development of future pack members and the tricks of life. Hatier witnessed several occasions where the alpha female taught her young how to play tug of war. He speculates that this game is a means of preparing them for the day when they are needed to disarticulate carcasses, in addition to giving pups a valuable chance to develop their coordination and strength. Pup classes are a microcosmic reflection of the overall pack. Soon after they are born, a dominance hierarchy is established within the litter. Through sheer brawn and battles for regurgitated food, dominant pups place themselves first in the pack's pecking order. No one knows for certain if a dominant pup retains its position throughout the rest of its life. By twelve weeks of age, the hierarchy is already in place.

Most of the time each member of the pack performs specific tasks that are vital to maintenance of the group. But there are, however, instances where some betas do little. These pack members are called "slouches" and may not be tolerated by their relatives. It can be a tough, tough world out there beyond the sanctity of the home range. Should they not attach themselves to their own pack, they may set out looking for a new territory to colonize. It is a high-risk gamble because the young adults are green in their hunting skills and generally inexperienced with evading predators, particularly those outside the park with guns.

In a disrupted environment where coyotes are killed by humans indiscriminately, the kind of established social order found in Yellowstone breaks down. There might be many coyotes in a certain area but they would be loosely affiliated and not loyal to packs. Correspondingly, the lack of packs could also result in the abandonment of territorial delineation. If the disruption is extreme, such as widespread indiscriminate killing by predator control programs, the cohesiveness of clans might be thrust into a state of chaotic flux.

The tight pack structures witnessed in Yellowstone, it turns out, may actually be anomalies. Far more common, at least outside of protected areas, are small groups of individuals that may, or may not, be immediate relatives. The advantages of forming a pack, besides protection, have to do with economizing energy when food might be scarce, and when several animals are needed to bring down large prey.

Biologist Jennifer Sheldon, in her book, *Wild Dogs: The Natural History of the Non-domestic Canidae*, sorts out the difference between the two types of lone coyotes. "Solitary residents are coyotes that have an established home range with no cohabiting pack. Nomads are single individuals with no site attachments, who range over large areas." Sheldon notes that lone coyotes may travel great distances, as much as hundreds of miles. Crabtree's work in both Washington and Yellowstone has corroborated Sheldon's findings.

Biologist Franz Camenzind conveyed the essence of this behavior on the National Elk Refuge in Wyoming where large groups of coyotes gathered to feed on carrion. "Aggregations were large (7-22 coyotes) ephemeral groups that displaced no social organization and were composed of winter migrants and nomadic coyotes mingling with resident pairs and packs," Camenzind wrote in "Behavioral Ecology of Coyotes."

During severe winters when food is scarce, rival coyote packs in the Lamar Valley interact with one another almost magnanimously and permit brief trespassing through foreign territories. If a large ungulate should die or fall prey to hunting, the spoils are shared. Members of the host pack gorge themselves until their bellies are filled. They may be succeeded at the carcass by members of an adjacent pack. Until the carcass is picked clean, wave after wave of visitors might assemble. Nearly two dozen different coyotes have

The Western rangelands have been a war zone for the coyote which has faced an arsenal of poisons, traps, bullets, and a host of other torturous horrors. Somehow, the coyote shows its resiliency by always bouncing back.

been seen staged around a single carcass with members of up to four packs present. The remarkable thing is the level of tolerance and non-antagonism, an element that is uncommon among rival wolf packs.

In this regard coyotes appear to be more socially advanced than wolves. "Physically traveling together [in packs] is only one of many parameters that you gauge how socially evolved a species is," Crabtree says. "There are a variety of behaviors with coyotes, especially at carcasses, that appear to function as means of avoiding fatal encounters. Whereas wolves tend to kill other wolves in scrapes over carcasses and in territorial conflicts, such killing in coyotes is absent." He describes a dance known as the "alligator gape" that coyotes perform when they are making a bid to eat at carcasses held by foes. The visitors get stiff-legged, arch their backs, protrude and lower their heads, bare their fangs and snarl. And yet, all the while, tails are tucked which is a classic signal of submission in canids worldwide. Crabtree interprets the ritualized jig as a statement that suggests: "I want access to the carcass and I'm damned hungry. I think you've had enough. Let me eat, but remember we don't have to spill each other's blood over this." Generally, the hosts comply.

In nature, some animals form partnerships to aid in the food gathering process. One of the most fascinating cooperative arrangements is that between coyotes and badgers. With the diets of both species centered on rodents, cornering the burrowing prey can sometimes be difficult, especially if there are several escape hatches present. Scientists studying both species of carnivores have documented cases where coyotes and badgers approach a complex of burrows together, with one animal starting to excavate a meal through the main entrance while the other waits at another hole to ambush the fleeing gopher or mouse.

"Badgers are not unfriendly toward coyotes," writes Hope Ryden in her book, *God's Dog*. "On the contrary, Indian legend has it that the two are companions, even soul mates. The Aztec word for badger is italcoyotl, meaning 'earth coyote,' and, according to the Navajos, Coyote and Badger called each other cousin."

Badgers aren't the only critters that assist scrounging coyotes. The wild dogs often are followed by ravens and other meat-pecking birds that descend upon a carcass once a kill is made. Likewise, coyotes have watched the aerial movements of ravens, vultures and other avians to locate carcasses or infirmed animals that are about to die. Occasionally, bird feathers are found next to the carcass site, but normally coyotes display tolerance toward their feathered fellow-denizens.

Hunting for prey with multiple routes of escape often can be difficult, but coyotes and badgers, though normally competitors, seem to temporarily tolerate each other while foraging. As the badger moves in to excavate ground squirrels or prairie dogs, the resourceful coyote waits intently outside an escape hatch to nab the fleeing rodent.

Early in their tenure, the alpha male and female from the Bison Peak pack are observed picking a carcass clean. The skills they honed here led to their reputation of being known elk killers.

Capitalizing on their own resourcefulness, the alpha pair of the Bison Peak pack had taught their colleagues every trick in the book. They were a model for how the predator-prey relationship is supposed to work in natural systems, and the researchers studying them had a special eye for the seemingly indomitable pack.

Love. Affection. Dependency. Loss. Humans turn to such nouns because they are easy, descriptive touchstones to get at feelings that cannot be quantified either empirically or simply by tacking on more adjectives. For wildlife biologists, trying to attribute such value-laden interpretations upon acts of animal behavior leaves a person in dangerous water. Anthropomorphism is said to have no place in objective science. University professors remind their students to stay away from it like the plague. But the fact is, humans are fundamentally biased organisms. We turn to our emotions to help us make sense of what happens in the world around us. We grow attached to animals that allow us to get to know them better. Denying those feelings is being dishonest.

Camenzind, the canid researcher and filmmaker who spent half a decade with coyotes on the National Elk Refuge, does not ascribe human characteristics to his subjects. But there is something about coyotes, he says, that makes them prone to anthropomorphism. "I've had animals that would tolerate my presence one on one, permitting me to walk with them for 40 to 50 yards. They will accept you if you accept them," he says. "I am continually amazed by how so damned incredibly clever they are. Just about any anecdote relating to coyotes I believe. Recently I was watching a group of five. There's no other way to describe it other than to say they were flat out having fun, playing and chasing each other. They were there in the wild, at home, and life was good for them. It was enjoyable to see that, to see any animal that at ease in its environment."

Crabtree echoes those sentiments and adds, "The similarities between the social and breeding systems of the coyote and humans are striking. Coyotes like humans, attempt to mate for life, are territorial, and build social units consisting of family members with parents, brothers and sisters helping to raise the young."

Love. Affection. Dependency. Loss. Again, these are not words that Bob Crabtree is comfortable contemplating when he describes the coyotes of

Lamar but he suggests they may have relevance within the context of the Bison Peak pack. As he tells me the tale of how No. 570 became the matriarch of Lamar Valley, I remember his musing during a stroll through her territory months earlier: "Unfortunately, the only certainty of being an alpha is that one day, you ultimately are destined to lose your position. You rise to power and you must fall but that can happen in any number of ways."

In a place like Yellowstone, despite all the steps that a researcher takes to safeguard his subjects against outside interference, inevitably they prove to be not good enough. Accidents will happen, and this one became a stone that was tossed into the coyote pond and sent ripples in every direction.

The average winter snowfall on Yellowstone's northern range is several feet. A portion of the Bison Peak territory was bisected by a highway that is plowed to make the Lamar Valley accessible for people passing between Mammoth Hot Springs and Cooke City, Montana. The plowed snow creates high headwalls on either side of the highway, in essence an inescapable gully that at times might be difficult to traverse, particularly if you are an animal two feet high. To scent-mark his territory and search for elk, the alpha male of the pack had to cross the road to pass from the Lamar River to the foothills girding Bison Peak. He made the pilgrimage perhaps once every evening. He was a formidable border enforcer and few members from rival packs were willing to mess with him, which gave the Bison Peak territory the air of an impenetrable fiefdom.

It was early morning, thirty degrees below the freezing mark, and Crabtree was driving to a trailhead for a snowshoe expedition and a full day spent with his subjects. Ahead of him a few miles down the road was the alpha male perched at the lip of the great roadway trench. Unbeknown to him or to Crabtree the park snowplow was barreling down from the other direction. At the point where the coyote stepped into the asphalt, there was a slight bend in the road. Crabtree met the snowplow driver who kept going with no indication of what happened. The alpha male had been struck and his body tossed atop the bank. "I got to him when the last life was being sucked from his lungs," Crabtree recalls. "This canid, from the standpoint of what wild coyotes represent, was the best of the best. I felt bad about the loss but I was more concerned about the effect it would have on 570 and the pack. They had been together nine years as a pair. The whole foundation of the pack was built on them as a team."

Older pups still attempt to beg a free meal from their father. Normally by this time of the year, they fend for themselves.

On three different occasions involving other wild coyotes, Crabtree has watched established packs disintegrate and social chaos ensue after the alpha males were killed. "I've watched some packs that were strong and cohesive the week before disappear almost overnight because the alpha female leaves, and without an alpha pair the pack can easily evaporate." He worried that 570's pack might follow the same course.

"She seemed shaken up, lost," Crabtree said. "She was going into her twelfth year, which is damned old. But she remained, and because of her the Bison Peak pack did not break up. It went from being patriarchal to matriarchal, leaving her at the top of the social hierarchy. But this is not to say there wasn't an impact. The loss of her mate still is felt in the Lamar Valley to this day."

Preceding page. One of the endearing qualities of coyotes to humans is that alpha pairs often mate and stay together for life. In Yellowstone, the alpha pair from the Bison Peak pack was nearly inseparable.

MEDICINE WOLF

THE IMAGES OF OLD MAN COYOTE, the Singing Trickster and the Medicine Wolf have been embedded in the human consciousness for a long, long time. On the slopes of buttes jutting from the prairie, in red, slick-rock canyons, and inside the sheltering walls of caves, vivid wildlife reflections created by ancient aboriginal artists have all featured the coyote. Native Americans and coyotes, the two are synonymous.

By looking at the entire scope of human interaction with coyotes, we find that people have been getting along with the canid predators for far longer than they have been damning their existence. For the last 400 years, the European model of intolerance may have prevailed, but for 10,000 years and possibly longer, coyotes have been, and still remain an important part of Native American identity.

It is fitting that the indigenous name for the animal, not its Latin descriptive, endures as our first word of reference to the species. Coyote (it

For thousands of years, native peoples have celebrated the coyote in a variety of art forms. Petroglyphs, such as this—some of which date back to ancient times—show that humans and coyotes have not always been adversaries.

can be pronounced as either kigh-ote or kigh-oh-tee) comes from the Nahuatls, a Mexican tribe that referred to the dog as "coyotl." When the Spanish conquistadors landed, they dropped the "l" and added an "e."

"No other animal of North America has by name so penetrated the American/English and Mexican/Spanish languages as the coyote," wrote J. Frank Dobie in his 1947 classic, *The Voice of the Coyote*. "Standing alone, singular and plural, prefixed and suffixed, Coyote is one of the commonest place names of Mexico and the Southwest."

The ancient Aztecs had ornamental symbols depicting coyote as the god of pleasure and hedonism. Indeed, when coyote is loose in the neighborhood he is a guest you need to keep an eye on. In some Native American villages, coyote became humans' best friend and was a replacement pet for the domestic dog.

George Horse Capture, a member of the Gros Ventre tribe, and a nationally recognized authority on native art and culture, says it is important to note that native peoples do not vaunt the coyote as the Creator incarnate but rather a mirror held up to humanity itself. "Depending upon the area of the country, the coyote is a very extraordinary animal. Indians are smart by recognizing him," Horse Capture says. "They look at animals and personify them in their myths and legends to explain the world. One of the most diverse and powerful is the coyote. To us, he is Old Man Coyote. I saw a wall-hanging once that speaks to the playfulness and seriousness to which the animal is regarded. The message on it said 'And on the sixth day, God created the earth . . . or was it coyote?' That summarizes it for me."

Horse Capture was hired to help coordinate exhibits and construction of the new National Museum of the American Indian, a branch of the Smithsonian. He says the coyote has contemporary relevance in the lives of Indians who feel disenfranchised and persecuted. "I always relate Indian people with coyotes," he said. "We're regarded in similar ways. We've been on the periphery and never fully accepted yet we have our own great abilities in our own world and we are beautiful and we thrive. We're everywhere and people can't get rid of us."

Coyotes in oral tradition are as much a part of Native American culture as the morals of Aesop or the personifications of the Greek and Roman gods are to western Europeans. These wild dogs represent a fundamental truth revealed through allegory. Dobie wrote in *The Voice of the Coyote*: "To many

Clever coyotes survive in any terrain.

tribes within its range, the coyote stood—and yet stands totteringly—as a god, more significant for cunning than for morals."

Just as there are two sides to any human's character, coyote is held up as a demi-god of interdependent opposites that produce interesting contradictions. At once, the coyote can be magnanimous and caring, the next minute he can turn cruel and sneaky. Old Man Coyote is capable of being a god, a person, and an animal all at once.

With coyote as the scheming, mischievous traveler, certain stories tell of how he created the world, brought fire to the people and caused grizzly bears to sleep in winter. There are literally hundreds of stories found within Native American legends that have been passed down among tribes throughout the ages.

Typical of how native people have deified the coyote's resourcefulness is the story of the singing trickster, whom the Sioux called Iktome. The legend has a kindredness with the teachings of Aesop: In the old times, a flock of ducks were flying across the prairie when they spied a traveler loaded down with belongings on his back. The migrating fowl asked the old man what he was doing, and he said he had a bundle of songs in his possession.

Begging him to sing a few notes, they gullibly agreed to enter a tipi the traveler had built. He told them to sing these words: Ishtogmus wachee po! Tuwa etowan kin. Ishtah ne sha kta! or loosely translated "Dance with your eyes closed! Whoever opens them will have red eyes forever!" All the while that the ducks danced with their eyes closed, the traveler (coyote, of course, in disguise) was quietly stuffing some members of the flock into his bag to eat them for supper. Finally, when the survivors discovered the trick, they fled, and upon opening their eyes, coyote's spell was exacted: their eyeballs turned red.

When we think about the link between humans and canids, realizing that there are only a few genetic differences between coyotes, wolves, and every breed of domesticated dog, it is no surprise that the place with some of the oldest fossil records of coyotes—New Mexico—is situated near some of the oldest archaeological records for permanent human inhabitation of North America. This is the region of the country where the notion of coyote as "the medicine wolf" was born and then spread to other tribes. Eventually, mountain men and trappers came face to face with the myth.

"Jim Bridger, Bill Williams, Old Rube, Sol Silver, Black Harris, Broken-Hand Fitzpatrick, Uncle Dick Wootten, Grizzly Hugh Glass, Blackfoot Smith and the other Mountain Men lifted hair, danced their own scalp

Contrary to their adaptability, wind and rain don't favor the coyotes' hunting activities. Environmental noises muffle the sounds of prey.

Playing and games are not only pup activities. These two adults pause to rest after a morning of fun.

dances, lived for months at a time on nothing but meat, sometimes eating liver for bread, and were, in fact 'the white savages of the West,'" wrote Dobie. "Most of them had strong faith in medicine wolves, called also medicine dogs [by native tribes]. Some early travelers reported these animals to be 'supernatural,' 'phantom wolves;' others described them as a species of feist-like [smaller] size resembling the jack rabbit. The medicine wolf was nothing else than the coyote."

Catholic missionary Father DeSmet, who traveled throughout the northern wilderness, wrote that indigenous peoples regarded coyotes "as a sort of Manitou. They watch its yelpings during the night, and the superstitious conjurers pretend to understand and interpret them. According to the loudness, frequency and other modifications of these yelpings, they interpret that either friends or foes approach," he suggested in portions of his journals and correspondence published as *Letters and Sketches* in 1843.

During trips to the Southwest, I, too, have gone camping in search of the mythical moon dog and never had trouble finding his sign. In the mesquite of southern New Mexico after a rain, I stayed on a game trail that brought me to a domed hill and the wash of a ravine. A coyote had been there. Fresh scat had pockets of fur. Amid the white, crystalline sugar-loafs of Great Sand Dunes National Monument a few years later, I followed tracks that converged upon the impressions where a snake had crawled and then was carried back to feed hungry pups. I was close, but no cigar.

A summer after that, I had my sleeping bag rolled out under the stars in the slick-rock country of the Colorado Plateau. There was a meteor shower, and a lone moon dog emerged from the brush and serenaded the night in sopranic cries. The animal finally materialized, silhouetted, as I imagined it would be.

Author Hope Ryden spent a few years studying coyotes and wrote about her research in National Geographic and in her book, *God's Dog*. She understands why the animal is revered by native peoples. "For ultimately my field studies revealed the coyote to be an animal indeed more wonderful, more beautiful, and more to be admired than all the logical reasons I have also tried to set forth to demonstrate why he is so vitally important to whole biotic communities," she wrote. "An older and wiser culture understood all of this when they spoke of him as 'God's dog.'"

George Horse Capture says he regularly hears stories from his son, a trapper, who claims he has never seen a smarter animal. Somehow, if a coyote narrowly escapes danger, it is able to pass along information to others and they avoid it.

Horse Capture points out that coyote is held up in Native American culture as an extension of the Creator endowed with special powers. "Some tribes put him near the top but he is never at the top because we reserve a special place for the one above. The coyote is the dean. He has so many traits to humans besides trickster. He is a supernatural being. If you go to a tribe, he'll always be woven into their origin stories. Sometimes he is a hero, sometimes not, sometimes he's the one who goes and sleeps with your wife. Foremost, he is there to help us poke fun at ourselves."

During an autumn hunting season along the picturesque Missouri Breaks on the high plains, Horse Capture was waiting for others to push game toward him. Skulking through the grass to a point within feet of him came a coyote. "I raised my gun and then I brought the barrel down. I raised it up again and decided that I had no reason to shoot such a handsome creature as this. I let the coyote go without firing a shot and I've never regretted it to this day."

The Blackfeet Indians have an expression: "Whoever shall fire upon a coyote or wolf, their barrel will never shoot straight again." For indigenous cultures in many parts of the New World, spiritual reverence is accorded to wild dogs because in the old times long ago, the wise, seasoned coyote and its larger brethren, the wolf, taught warriors how to hunt buffalo. As a token of thanks, humans are expected to repay the coyote's gift of wisdom with kindness. Blackfeet tribal elders say respect for canids brings "good medicine." Killing the four-legged predators can be a prelude to bad luck.

Held up as a demi-god in certain native religions, the coyote is said to possess mystical power. Whoever shoots their gun at a coyote or wolf, one legend says, will have a barrel that never shoots straight again.

FOR MUCH OF THE TWENTIETH CENTURY, American taxpayers have funded a quiet, but unsettling war against the coyote. While there are no exact estimates on how much money has been spent attempting to annihilate coyote populations and other predators, were we to consider all of the funding paid by the federal government, state agencies, the livestock industry and individuals, the figure, conservatively, could be placed at hundreds of millions of dollars.

While this expense by itself is enormous, studies show that the coyote killing fields on private and public lands have taken an even greater ecological toll. The irony is that in spite of the weapons aimed at *Canis latrans*, the species has emerged not only more abundant and widespread but as a direct result of human efforts, mankind has created a predator better able to withstand persecution. The era of the "Super Coyote" has arrived.

By playing God with predator populations, the federal government's

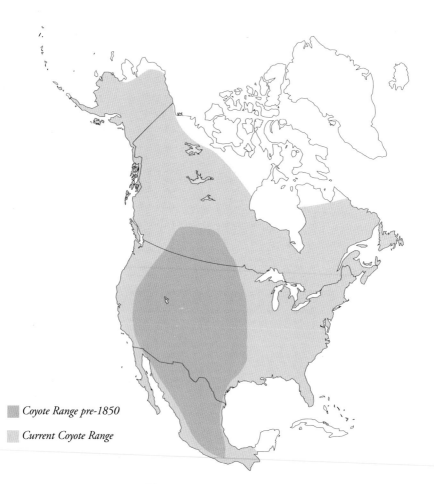

Coyote Range pre-1850

Current Coyote Range

Animal Damage Control Unit (an agency within the U.S. Department of Agriculture) has emerged as a lightning rod in the debate over coyotes and how society will treat them in the next century.

Few people understand the impact of ADC better than Dick Randall, a former employee who spent part of his youth trying to eradicate coyotes from western rangelands in the heart of cattle and sheep country. Although the World War II veteran is labeled a turncoat for trying to reform the government's policy on predators and shape public attitudes toward coyotes, there was a time in his life when Randall was the coyote's worst nightmare.

Throughout the modern history of wildlife management in North America, no other creature—not even the wolf—has been more maligned and demonized than the coyote. Trapped, shot, burned, drowned, gassed, snared, poisoned, blown up with dynamite, dismembered, used as target practice, strangled . . . there isn't a torturous tool that has escaped the coyote's hide.

Before he retired from his job as a depredation technician with the U.S. Fish and Wildlife Service, Randall's aim apparently did not fall under the spell of the Blackfeet's tribal legends that say guns shot at coyotes never shoot straight again. Randall was, in fact, a deadeye when it came to laying crosshairs on the body of a coyote from several hundred feet in the sky. He also was proficient at lacing the carcasses of cattle and sheep with deadly doses of poison to kill coyotes and dragging whining pups from their natal dens.

For ten years, Randall flew as an aerial gunner and complemented his job with a trap and poison line on the ground. "I was good at it," he says, "but just because you're good at something doesn't make it right." During one winter of coyote control in the sagebrush-covered plains around Rock Springs, Wyoming, Randall achieved notorious distinction: he set a record for most coyotes killed in a single day when he shot forty-two animals in six hours; the same month, he broke another record with 230 casualties. "It had been a terrible winter with lots of snow, leaving the gullies filled-in so there was no place to escape or hide," he says. "If you flew 500 feet high you could spot a coyote a mile away and he had no place to go. That winter we had three aircraft killing coyotes and we decimated the population. It was interesting because the next spring ranchers were reporting the same amount of livestock losses and we knew that couldn't be. When we went to look for dens to kill the pups, it was difficult even to find coyote tracks. We had done our job but

some of the ranchers weren't being honest," he says.

The insatiable hatred toward coyotes, the hints that perhaps livestock losses had been exaggerated, and the fact that killing coyotes was not always solving the predation, planted that seed of doubt in Randall's mind toward his job.

The year after he had amassed record tallies of coyote carcasses, and following the second airplane crash of his career, he retired with a citation for excellent service. The only thing was, Randall realized that killing the thousands of coyotes may have done far more harm to ranchers than good. How could this be?

"Intensive human control of coyotes can have exactly the opposite effect of what you hope to achieve," Randall told me. "Let's say you have seriously reduced the number of coyotes from a given area. When Mother Nature sees this, she whispers in the ears of the surviving animals and says 'This ecosystem is void of coyotes so get busy and reproduce.' Coyotes are able to respond to pressure," he adds. "As soon as you start culling the population, the pup count goes up and the killing goes up because the adult animals have to feed the pups. In an area of intensive predator control you may have six to eight pups per litter, while in Yellowstone, where there is no control, the litter sizes may be half that. Similarly, with exploited populations you find that between fifty and sixty percent of the yearlings breed, while in places like Yellowstone the animals might wait until they are two years old before they produce young."

As long as non-Indians have inhabited the North American continent, a large segment of society has maintained a negative attitude toward predators that were imported from Europe and fueled by such stories as Little Red Riding Hood and the Three Little Pigs. Generally, animals that ate other animals were considered "bad" and those that did not compete with man or provide food for the table were considered "good." The same logic that produced fairy tales was extended not just to wolves but to any wild, meat-eating animal.

At first, it was big carnivores that were targeted with eradication—wolves, mountain lions, and grizzlies. The livestock industry officially declared war against predators in 1907, but it would take eight years until Congress committed U.S. taxpayers to the effort. It earmarked $125,000 to the then-Bureau of Biological Survey to eradicate wolves. Initially, coyotes

As snow deepens, capturing voles and other rodents under the snow becomes increasingly difficult, forcing coyotes to consider other food sources, including scavenging on carrion.

were tolerated by ranchers because they ate a lot of the mice, prairie dogs, and other rodents that survived on the same grasslands grazed by cattle and sheep.

But by the middle of the twentieth century when the value of large predators was finally recognized by scientists, and most of the bear and wolf populations had already been wiped out, coyotes became the top bane of ranchers. The more that coyotes were killed, it seemed, the greater the problem became with them eating sheep. No public or private land was spared of the predator purge, not even national parks. Between 1904 and 1935, 121 mountain lions, 132 wolves and 4,352 coyotes were killed in Yellowstone National Park alone, which illustrates the coyote's relative abundance in nature compared to other large predators.

In 1940, biologist Adolph Murie published his seminal report "Ecology of the Coyote In Yellowstone" which helped convince the government to abandon predator control in national parks. Murie was supported by his brother, Olaus, who spent several decades working for the Biological Survey, and against enormous political pressure exerted by the livestock industry, spoke out against the government's campaign to eradicate the carnivores at the top of the food chain. Olaus Murie, recognized as one of the finest American naturalists in the history of this country, went specifically to the defense of coyotes, recognizing their intrinsic value: "How can one speak adequately of this legendary little wolf except in terms of poetry?" Murie asked. "The coyote has woven himself into the folklore of the original Americans of the West. They recognized in him an important fellow-being, one that shared with them the dangers, pitfalls, and joys, the struggles and satisfactions of primitive life."

There are many who view the Murie brothers as prophets because more than half a century ago, they correctly ascertained that humans, by also eradicating the things that coyotes like to eat—rodents (because they compete for grass with cattle)—had actually created a predator that, in some cases, had no alternative but to turn on sheep and young cattle.

Ironically, just as the last wolves were disappearing from the West, the modern version of the Animal Damage Control Unit, on March 2, 1931, officially came into being. For years it was administered by the U.S. Fish and Wildlife Service but is managed today by the U.S. Department of Agriculture's Animal and Plant Health Inspection Service (APHIS). At the behest of ranchers, ADC had made coyotes Public Enemy Number One.

Depot agent Harris (left) and trapper Joe Holland are shown here with coyote and beaver furs at Homer T. Goodell's elevator. Courtesy Montana Historical Society.

Millions of coyotes were killed in the decades after World War II when the government picked up its attack on the species.

In addition to traps and aerial hunts, the government recruited other deadly weapons: super-lethal poisons called "predacides." Among them was Compound 1080 that William Ruckelshaus, former administrator of the federal Environmental Protection Agency, called "one of the most dangerous [toxins] known to man." During the 1960s, ADC set out hundreds of thousands of meat baits containing Compound 1080 to kill coyotes, and 1.3 million pounds of grain laced with 1080 to kill the rodents coyotes would normally eat. Also, over seven million strychnine tallow pellets were dispersed over western rangelands.

There was a problem, however, that apparently was overlooked: When you spread that much poison out on a landscape, it enters and flows through ecosystems unpredictably and in insidious ways. A coyote might perish as planned but what happens after it dies? Perhaps a golden eagle swoops down and tugs at its contaminated flesh. If it doesn't die, it might not be able to reproduce, and if the poison does destroy it, there might be a raven or buzzard recycling the deadly compound. And it wasn't just coyotes stumbling across poisons but pets were ingesting the strychnine and Compound 1080 and dying, too.

Back in 1964, the federal government launched an investigation into ADC and asked Starker Leopold to head a committee preparing a report. "It is the unanimous opinion of this board that control as actually practiced today is considerably in excess of the amount that can be justified in terms of total public interest," Leopold wrote. "In short, the federal predator and rodent control program is to a considerable degree shaped and designed by those who feel they are suffering damage from wildlife. There is no mechanism to assure that the positive social values of wildlife are given any weight in decision making nor that control, when it is undertaken, will be limited to minimal needs."

In 1972, President Nixon outlawed the use of these super-lethal poisons, but when Ronald Reagan came into office in 1981, the ban was reversed. While environmentalists have pushed for a permanent ban on poisons, law enforcement officials say there are many illegal agents that are being used. "We've known it's been going on for years. We just didn't know the level of it until we got into it," said Galen Buterbaugh, who oversaw the Fish and

In unprotected areas outside of places like Yellowstone National Park, coyotes are susceptible to high levels of human-caused mortality. Notice the .22-caliber hole in the cranium.

The leghold trap is the most common method of control and also may be among the most controversial. Scientists argue that indiscriminate, wide-spread trapping is biologically counter-productive. Data indicates that lower adult population levels ensures the survival of pups in the litter which in turn may lead to more sheep killing.

Wildlife Service's Rocky Mountain region. A 1991 sting operation, for example, led to the arrests of people who were involved with poisoning thousands of bald and golden eagles with enough toxins "to kill every man, woman, child and mammal in the western U.S." Among several investigations that the U.S. Fish and Wildlife Service has conducted relating to lethal control of coyotes, it also raided what was considered to be a black-market depot for banned chemicals, but equally as troubling to opponents of ADC is the materials that still can be purchased legally.

Among some of the poisons and devices employed to kill coyotes over the past few years have been zinc phosphide, strychnine, and the equally notorious "coyote getters" that send tablets of M-44 sodium cyanide into the mouth of an animal that approaches government baits.

Last, but not least, the government has come under fire for advocating the practice of "denning" where coyote mothers are tracked to their dens and then the pups are pulled out and killed.

A basis for the livestock industry's persecution of coyotes is the belief that left uncontrolled, predator populations will increase ad infinitum. It is a familiar assumption that fans the flames of hysteria, though it ignores the basic laws of nature, the first being that a predator population fluctuates in accordance with its prey base. In other words, coyotes appear to be self-regulating. The premise was shown by Murie's study of unexploited coyote populations in Yellowstone during the late 1930s, findings which were subsequently reaffirmed by Bob Crabtree in Yellowstone fifty years later. "Apparently," Murie concluded, "the Yellowstone coyote population does not increase indefinitely. Facts enumerated show that the population level is kept down by disease, possibly in some cases by starvation, and that this species is subject to natural controls."

Franz Camenzind said that ADC's and the livestock industry's war on coyotes has been a tragic tale of "overkill." "Probably the most universally held misconception is that coyotes have to be controlled or they will destroy everything around them. It is a complete and utter fallacy," says Camenzind, who spent much of the 1970s in the field with coyotes on the National Elk Refuge and during the 80s and 90s making wildlife films. "They [coyotes] control themselves long before they affect prey populations. People have this consuming passion to control things and suggest that if you don't there will be a coyote lurking behind every clump of sagebrush. The logic is that if you

keep killing them you will keep them under your thumb. The irony is that coyotes have such reproductive potential that it doesn't work, unless of course you go out and try to kill every last one of them. While some people may favor that approach, society isn't willing to put up with it anymore."

Of course, there are cases where ranchers have attempted to be tolerant of coyotes only to be run out of business by mounting losses. Coyotes kill, sheep die. You can't blame a family working toward retirement or saving to send a child to college for wanting to protect their investment against an animal that is perceived as the enemy. Where coyotes are unrelenting, most sides usually agree: The ADC is needed—to an extent.

Jim Till, a former ADC trapper, researcher and respected biologist with the U.S. Fish and Wildlife Service who helps the government relocate marauding wolves in Montana, says the issue is not clear-cut. The name Animal Damage Control speaks to the mandate of the agency. It is to control damage caused by wild animals on domestic livestock. While he believes that reforming the agency is necessary and in some ways, long overdue, after seeing the impact of coyotes firsthand in the field, he also understands the need to keep it going. Till says that ADC is already making a shift from all-out population control into the realm of finding only sheep-killing coyotes and eliminating them from the population. Even that, however, sometimes results in innocent coyotes being killed.

"It really isn't so simple as going to an area where sheep have died and putting out a little predatory brush fire," Till says. "I don't like this analogy but it seems to work. If you spray an insecticide in your back yard to kill insects which bother you, you're still going to kill other bugs that die just from being there. Innocent coyotes will die from control measures, that's a given."

Till undertook a novel research project: He compared the effectiveness of eradicating all coyotes from a given area where sheep were killed versus simply removing the pups. Ranchers were losing lambs in the spring and early summer because adult canids were pressed to keep hunting prey to feed their young. Once the pups were removed, the adults often turned back to rabbits and mice in their foraging because the pressure to produce large quantities of food was no longer there. As it turned out, lamb loss was reduced by about the same amount with either method.

"Some people are opposed to killing coyotes no matter how justified it

Siesta time for active pups under the natural roof of a juniper in Montana.

might seem to be, and there are staunch people on the other side who say the livestock industry will go under without ADC," Till says. "Control of coyotes using ADC has become almost like a moral or religious issue, and nearly as polarized as the wolf issue."

It is easy to become self righteous about the perceived atrocities of coyote control while living in an urban area (where most Americans are these days), eating a vegetarian dinner and watching a pretty documentary about coyotes on television. But we all must check our closets and think of the wool we wear. Did those fibers come from an American producer? If they did, then there's a chance the sheep that yielded the wool were protected in part by taxpayer dollars going to work by killing coyotes or keeping the animals at bay.

The need for an agency like ADC becomes desperate if you are a rural wool producer with all of your future investments counted not in stock certificates but heads of sheep. You can be as sympathetic to coyotes all you want but if the predators are eating three or four ewes or lambs each night, the economic impact adds up quickly. By far, the greatest number of conflicts between ranchers and coyotes occurs in the seventeen states that comprise the West. In 1992, roughly 100,000 coyotes were killed west of the Mississippi compared to 130 east of the river. The same year, coyotes allegedly caused almost $2 million in damages compared to $3.8 million the year earlier and $4.6 million in 1990.

There are many skeptics who say the financial tolls are inflated and that coyotes have become a scapegoat, being blamed for losses caused by other predators, natural death or simply poor flock and herd management. Outcries from citizens over the destruction of predators has brought discussion of non-lethal alternatives to the forefront in recent decades. A spectrum of environmental groups, including the Humane Society of the United States, Defenders of Wildlife, Wildlife Damage Review, the Predator Project, the Biodiversity Legal Foundation, the National Wildlife Federation, the Animal Welfare Institute, the Sierra Club and the Fund for Animals, all have pressed the government to take new approaches. So far, the most innovative solutions—and ideas that now are being universally adopted by farmers and ranchers—have come from the private sector. "I think people will tell you, the only great innovation in the sheep industry over the last few years has been electric fencing of flocks, and guard dogs," says Ray Coppinger, who proved instrumental in the implementation of the latter.

In 1976, Coppinger, who is professor of biology at Hampshire College in Amherst, Massachusetts, was turned loose on a pioneering experiment. He had heard stories about shepherds deploying special canid deterrents in Europe that curtailed wolf predation. Coppinger gathered as much information as he could and brought several breeds of dogs into his American study.

Formerly a successful sled dog racer and breeder, he worked with three premier recruits—the Anatolian shepherd from Turkey, the Shar Planinetz from Yugoslavia, and the Maremma from Italy—and then enlisted farmers who had coyote problems. Guard dogs work this way: released into a flock when they are between four and sixteen weeks old, they bond with the sheep and, similar to an alpha coyote protecting his clan, the guard dogs serve the same function. Coyote kills dropped off dramatically. When the study ended fifteen years later, some 1,400 dogs were in the field and the number has grown ever since. In Utah, ninety percent of the flocks are protected by these proven breeds. A panacea? Not at all. Merely an option.

The family of rancher Teddy Thompson has been running flocks of sheep to grasslands in the Absaroka Mountains of Montana since 1921 and over the years they've had coyotes and grizzlies prey upon the ewes. Some nights, fourteen or fifteen sheep were taken by coyotes and it prompted Thompson to take action. He imported a pair of Turkish Akbash guard dogs and the losses dropped to nearly nothing. The large, aggressive, domestic canids will actually chase down coyotes and kill them. Still, Thompson says if environmentalists are successful in dismantling the ADC a backlash will be felt across the West. Vigilante ranchers will take matters into their own hands and start putting poison out again, even though it is illegal. "In the old days when poison was used there were a lot of predators and other animals that didn't need to die," he says. "The same thing will happen again if you do away with the ADC and it won't be pleasant on the animal kingdom. ADC is needed because we can't change the coyote and turn it into a vegetable lover. Coyotes are a predator. Meat-eating animals prey on things that are easy to catch. They're opportunists. The story is that they always kill the weak and the sick but that's a fallacy. They'll kill your best lambs."

Guard dogs, guard llamas, electric fencing, and artificial noise repellents all work to a point as coyote deterrents, but in areas where predation is a persistent problem it requires vigilance from wool growers to mix their methods. Coyotes are smart; it doesn't take them long to figure out how to run a

European breeds of guard dogs have become a popular and effective means of non-lethal coyote control in areas where flocks of sheep graze. The dogs bond with the ewes and defend them against wild canid predators.

predictable gauntlet; they need to be confronted by an element of surprise.

"I encourage, and have for years, that people take on guard dogs because they can cut your losses to nothing," says Bob Gilbert, secretary-treasurer of the Montana Woolgrowers Association. Gilbert says the approach, however, is not foolproof. Coyotes have learned in certain situations to circumvent the dog by having one group of predators serve as decoys while another group gets into the unattended lambs and ewes. "Guard dogs work where you have your sheep in small pastures but in open areas of the West there's just too much real estate to cover."

Coyotes, he claims, have inflicted annual losses of fifteen percent on some flocks and completely driven wool growers out of business. He says that if it hadn't been for ADC, many other sheep producers would have gone broke. Gilbert doesn't vilify coyotes. He just decries the government for slashing its predator control budget. "The coyote is a meat eater," he says. "He is not going to subsist on a diet of lettuce and grass. They are cunning and are still around despite the efforts of man. But they are also a predator that causes tremendous financial losses that affect families and real people. I suspect that if we abandon control of coyotes you'll see ranchers going under. As we do away with people in rural America, we're going to lose a philosophy, a way of thinking, that really made the West."

Within scientific circles, there is growing attention being paid to a theory known as the "predator-control paradox" which asserts that the more traps, poison and bullets thrown at coyotes, the more likely that conflicts with livestock will continue.

Could it be that coyote populations which are not persecuted by bullets, traps, and poisons actually represent less of a threat to domestic livestock? According to Bob Crabtree, who has observed coyotes in Washington, California and Yellowstone National Park, the answer appears to be "yes."

Crabtree has been in the process of synthesizing information gleaned from studies of coyotes to develop demographic modeling which will be used to predict what effect the reintroduction of wolves to Yellowstone will have on their smaller cousins.

Crabtree says that livestock grazing, when combined with campaigns to control coyote numbers, may actually spur more killing of sheep. The reason is that when dominant leaders of coyote packs are killed by federal trappers, the social system of coyotes is radically disrupted.

As a result, aggressive sub-adult animals strike out to establish their own territories, producing even more coyotes than existed before the first animals were destroyed. Because livestock practices also result in fewer rabbits and voles inhabiting open rangeland, coyotes are then forced to take easier, unnatural prey in order to survive. Crabtree says persecution by humans has produced "the Super Coyote." "As hard to believe as this may sound, wild coyotes which have not been persecuted actually have an aversion to animals they are not used to, including sheep," Crabtree notes. "If given an abundance of natural prey, in many cases they would probably leave livestock alone. There are a few examples where coyotes and sheep were placed in large enclosures and the sheep were left alone."

Till and other trappers have witnessed many instances where coyotes instilled with natural foraging skills have co-existed with sheep flocks over generations of animals. Often, however, coyotes that have lost the ability to scavenge for natural foods become the culprits in attacks on sheep. The instinct to naturally forage is taught to yearlings and pups by the alpha males and females. Again, coyote pack structures are based on a pecking order and stability. Start removing the members of a pack in a piecemeal, haphazard fashion and you've created a recipe for chaos. By killing off the alpha pair, there may suddenly be multiple animals vying for dominance, and those which do not emerge as leaders in the new pack may strike out on their own and turn to sheep for sustenance because they have no natural foraging skills.

"The rancher thinks that if coyotes are killing sheep, then by killing coyotes you have less predation, but that isn't necessarily the way it works," Crabtree says. "Livestock producers have been led to think of predator control only in a straight line but nature is dynamic in its function. It doesn't always react the way we think it will."

At the same time, he adds, environmentalists need to be sympathetic to what sheep and cattle producers are confronting. "Many environmentalists are not willing to perceive the coyote as a real economic threat to the livestock industry, which it is," he said.

Rather than viewing coyotes solely as pests, some environmentalists are suggesting that the coyote's beneficial role be accounted for in the landscape. The key is changing the cultural bias against coyotes by educating traditional land users. Tom Skeele, who heads an organization called the Predator Project, is pressing for reform of the way the government conducts predator

Pups begin playing "pack." Instinctively, the pups are already exhibiting many behaviors of their adulthood such as greeting and submission. Dominant pups have already established a social order in the litter.

control. "I think the future of predator control is dependent largely upon our ability to get away from looking at wildlife as being either good or bad but simply to respect its higher purpose, and I don't mean its purpose for humans," Skeele says. "We as a society need to get away from anthropocentric attitudes toward coyotes and shift into biocentric views."

However, in the America of the 1990s, Randall says sadly, little of the enmity is showing any signs of abating. At so-called "varmint shoots" in nearly every one of the Western states, contests are held that offer cash prizes for the person who can blast the most coyotes in an afternoon. When a wandering gray wolf was accidentally shot outside of Yellowstone National Park—the first wolf confirmed there in decades—the hunter who killed the animal said he had mistaken the wolf for a coyote. When asked why he wanted to kill the coyote unprovoked, he replied, "Because that is what you do when you see a coyote. You shoot it."

Despite attempts by humans to subdue coyotes, the animals have proved their resiliency. Every time a coyote survives a purge, it gets smarter and passes along the knowledge to future generations. The trick to living with coyotes may not be forcing them to be lesser predators, but convincing humans to stop tinkering with the system. After all, changing the way we treat coyotes may mean changing the way we think about ourselves.

While coyotes kill their share of mule deer, this time the tables have turned for Canis latrans. *Once detected, the coyote was found sleeping in a prairie dog town and run off by a herd of deer.*

GENERATIONS BEYOND

WATER FALLING FROM THE SKY always foreshadows an appearance of fresh tracks. Moisture, in whatever form, prepares the earth's surface to recite a story. The last chapter in the life of a coyote matriarch commences with a blizzard. There is now a matrix of canid pawprints sewing their seams across the Lamar Valley, connecting generations of coyotes from the past with those in the present, but leaving only impressions that will fade and one day be erased by a coming canid insurgency.

Current tracks cross snowdrifts gathering since mid November.

During the last few weeks, the Bison Peak pack had shown signs of restructuring. The tell-tale appearance of an alpha pair tenaciously scent-marking the perimeter of the territory was gone. The aging matriarch had stayed solo. An act of grief? Or was it the self-knowledge that her own end was near?

Coyote tracks are one of many clues that help humans identify them on the landscape. Although nearly impossible to prove, Crabtree strongly suspects that coyotes can identify individual animals from smells in tracks, urine and feces.

A dowager now, she inherited a prime spread of the natural world comparable in size to New York City's Central Park, and was bequeathed the task of looking after a pack that she and her deceased mate assembled over the span of a full decade. Researchers kept her in their spotting scopes as much as they could because they knew she served as a catalyst for either solidifying the pack or allowing it to fracture. The focus on her behavior had poignancy. Not only did she seem to be slightly traumatized by the loss of her lifetime companion, but she was approaching her tenth breeding season, a feat unrivaled within hundreds of miles and perhaps over distances greater still.

Plump and well coated, she was in remarkable condition, nonetheless, bolstered by an autumn windfall of grasshoppers, beetles, and voles. According to Bob Crabtree, she demonstrated no hints of slowing down, even though she was the equivalent of a spry senior citizen. "I will always be in awe of her, wondering if it was her superior genetic makeup or the incredible piece of real estate that the Bison Peak pack decided to settle on and defend," he says. "Probably, it is a fortuitous melding of both. She was amazingly fit and had the spring of a coyote half her age. You can recognize a coyote that is getting older and frail. She defied her years."

The oldest known coyote on record was a sixteen-year-old documented in Colorado. The matriarch of Lamar was just three years shy of reaching that milestone but she already had surpassed the average life expectancy of a coyote by eight years. Together with the alpha male, they had turned out dozens of coyotes with their genetic blueprints and were a model for canid vitality.

The previous twelve months, however, raised doubts among Crabtree's researchers about her ability to persist. When the alpha male perished, it happened three weeks before the breeding season. The pair was already well along in the process would lead toward her impregnation. And she seemed affixed to the instinctive routine. The loss of her mate caused definite anomalies in her behavior.

Initially, she had withdrawn from the pack except to defend the territory from interlopers. In the void that was created, a beta son began vying for his father's former position. This powerful coyote, No. 740, had led several hunts that yielded elk kills by using skills he acquired from the alpha pair. No. 740 was also the animal that spooked Crabtree over a trap and bit through his hand a year earlier.

Because coyotes mate for life, the loss of a mate can be disturbing in many ways.

Young pups need sleep as well as a healthy diet to survive.

Although Crabtree does not know which coyote, exactly, was responsible, the alpha female of Lamar became pregnant again. "We suspect it was 740 but there were a few other males around here trying to mount her," he says. "He was definitely attempting to become the next alpha male. When an alpha male dies or can no longer hold his position in the hierarchy, often the female will choose a successor. But she apparently decided not to pair up again in a permanent bond. Instead, she exhibited dominant behavior and he became submissive even though he was much larger. What's interesting is that in every other aspect except breeding, he was filling the alpha role by being a territorial enforcer and elk killer."

The previous year a rare double litter greeted the Bison Peak clan. Besides the eleven adult pack members, the bitch and one of her beta daughters each gave birth to a brood, which raised the census to twenty-three individuals! A banner year. As fate would have it, it would be the last direct genetic contribution the alpha male and female together made to their kin.

When spring came, the bitch appeared inattentive and was spotted on only a handful of occasions around the natal den. Only one or two pups were observed that year, and it is doubtful they survived. It was the first time that Crabtree had ever witnessed her fail. His concerns, he realizes now, were premature because the matriarch, while remaining solitary, returned to her former behavioral patterns and entered the fall in robust condition.

It had been a glorious season of transition heading into the winter. A golden glow beset the aspens that flanked the river. Elk had returned with their jazz resonating from mountain to mountain. Crabtree had now been following her for four years and he felt as though he were an extended member of the tribe. He knew the individual markings for a hundred coyotes stretching for miles in any direction, and he took personal pride in being able to point out sons and daughters of the female that had drifted away from Bison Peak, only to assume new roles as alpha members in adjacent packs. The blood lines of 570 flowed deep and strong through packs that adjoined the Lamar River. Apparently, nativity transcended allegiances to territories.

One day, while scoping the valley from a promenade beneath a Douglas fir, Crabtree saw a single coyote, Number 720, come loping toward the edge of the Bison Peak sanctuary. Instead of sounding a challenging bark, the beta sentry wagged its tail and allowed the visitor to pass. Normally, an invader would warrant a fang-to-fang growling match from 740 but the two seemed

to recognize each other. A beta in the adjacent Amethyst pack, 720 had helped serve her mother as nursemaid in previous pupping seasons. Crabtree strongly suspected that 720 may have been the offspring of a onetime Bison Peak beta that moved off and became the alpha in an adjacent pack. It shows the influence that the Bison Peak pack had on the Lamar Valley and led to several similar encounters between adjacent packs. Such reunions marked the last year of social convergence between the matriarch and her lineage because the end of a reign was coming to pass. Howling was starting to be heard off in the distance, but not of her kind. She would never meet a wild wolf, although for a long time she herself had been functioning like one.

TIME OF CHANGE

Although Western stockmen decry the return of wolves in wildlands in the lower forty-eight states, the large animal may actually help them resolve their so-called coyote problem. This hunch of scientists has to do with nutritional needs of predators. The arithmetic works like this: Remember the natural proportion that exists between foxes, coyotes, and wolves, how each species is roughly one-third as large as the next. Crabtree says similar proportion applies to territories, only to a much greater extent.

The size of a territory is determined by the amount of land necessary to produce food that will sustain the pack. Foxes are not pack animals but the individual range for a fox is about one-tenth as large as a coyote's, and in turn, a coyote's home range is one-tenth the size of a wolf's, roughly speaking. Over the same amount of terrain as is necessary to sustain a single wolf pack, you might find between ten and twelve coyote packs numbering seven members each. There might be eight wolves in the same space as eighty coyotes. Eighty coyotes consume far more than eight wolves, perhaps as much as four times more. Bring back wolves and in theory, at least, it is possible that ungulate and livestock populations would be under less pressure as a food source.

That scenario does not bode well for coyotes. And there are parts of the continent where coyotes could produce a net gain for other species, particularly waterfowl in the duck factories of the North American heartland. It has been shown in nature that coyote numbers fade when wolves are present, and fox numbers fade when there are coyotes present. One of the threats to duck populations in the prairie potholes of western Canada is the fact that nest

Coyotes use frozen lakes to explore islands which normally are inaccessible to them in the warm summer months. They have been known to chase ungulates such as deer and elk onto the ice where capture is almost assured.

sites, year after year, suffer from heavy predation by red foxes. These smallest of the canids have fared remarkably well around agriculture and cultivated prairies. Crabtree suspects that coyotes would dilute high concentrations of foxes and perhaps have an effect on other wildfowl predators such as skunks and raccoons. Thus, there might be more downies successfully fledging and filling the flyways every autumn. Coyotes themselves, then, can be a natural form of "predation" control, as Crabtree puts it.

"For so many years the coyote has been condemned as an adversary to man but we haven't taken the time to explore how it might be an asset to goals we want to achieve," Crabtree says. "I'm not suggesting that the concerns of ranchers and farmers are invalid because coyotes can have a serious impact on livestock. But maybe, just maybe, we have been approaching the coyote the wrong way. In some areas, the best strategy might be to simply leave it alone."

That's exactly what he had done with the coyotes of Lamar. Leave them alone. Let them behave like real coyotes do.

END OF A REIGN

Words like grieving or mourning do not enter into Crabtree's vocabulary because they are subjective and leading. Nature can be gentle but there is a side of it that has sharp teeth. Still, he wonders if the Lamar matriarch knew what was lying ahead. "It is not my role to read her mind," he said. "And even if I were to try, I would inevitably be wrong."

Following her routine, the alpha female set out alone. She returned to scent-marking the perimeter on cool December mornings to ward off invading coyotes that were interested in the pack's abundant rodent populations. She crossed the same road where her mate was mortally wounded by the snowplow and wound her way across the base of the high domed hills. She was securing the base of her pack's namesake peak.

Crabtree had left the park and returned to his home in Bozeman where he was compiling data that he had gleaned from his study for articles in scientific journals. On an afternoon a few days before Christmas, the telephone rang—it was Scott Grothe, one of Crabtree's research assistants.

"We haven't seen 570 in several days," Grothe said.

"She might just be out of reach. Can you draw a fix on her location?"

Crabtree asked.

"We're getting a shallow reading on her and it seems to be coming from the base of Bison Peak. But I should tell you, Bob, the mark on her position hasn't moved or changed in quite awhile. What should we do?"

"Wait until the morning to ski up there and carefully comb the area. See what you can make of it," Crabtree suggested. "Let me know the first thing you find."

Crabtree had a strong suspicion as to her fate. He waited to let the facts confirm his intuition. The search party led by Grothe glided on cross country skis a couple of miles to the eastern-most edge of the Bison Peak territory. Although the beep became stronger, there was no sign of 570. The terrain had become progressively rugged, accented by precipitous slopes bending down the mountain. Days earlier, another team of researchers, this one focusing on mountain lions, had keyed in on a set of cat tracks in the same vicinity. Three years almost to the same day in the same general area, an alpha female from the adjacent Druid Peak pack out on scent patrol had been killed by a cat. Mountain lions are the most dangerous predators for coyotes on Yellowstone's northern range. Crabtree calculates that lions kill five to ten percent of the coyote population each year.

Kerry Murphy, the biologist who coordinated a novel study of Yellowstone cougars with noted felid researcher Maurice Hornocker, had witnessed coyotes driving a lion off of kills, but he also found evidence of nearly two dozen head-crunched coyote skulls in his back-country travels. One on one, a single coyote is no match for the lithe agility and power of a cat. Typically, the prime method of cat predation is lying in ambush.

570's collar was still working but there was no body in sight. Crabtree surmises that the end of the matriarch was expedited swiftly. "You can usually tell if a coyote was killed by another predator by the condition of its radio collar," Crabtree said solemnly, holding 570's collar in his hand. "This one was clean and odorless. It seemed like the work of a mountain lion, a very clean kill and consumption." He says there is a single consolation: "If there had been any coyote capable of defending herself in a brush with a mountain lion, it was the bitch of Bison Peak, but she's normally quite aware of everything surrounding her. I would bet that she never saw it coming."

For the first time since 1982, the Bison Peak territory was left without an experienced leader and it might take years before the kind of stability galvanized by the bitch and her mate is restored. These short-term power struggles, however, are nothing compared to what is looming on the horizon. The tenure of the Bison Peak pack's next generation is destined to be only ephemeral, fleeting, a thought that is bittersweet to those who documented its primacy.

A changing of the guard is underway and a circle that began in the 1920s is about to complete another loop. In 1995, wolves returned to Yellowstone following a seven decade absence. Fourteen Canadian animals were captured and placed in holding pens before being turned loose to their new home. Their release sites happened to be the Lamar Valley, an area that was chosen for the same reason that it has been eminently productive coyote habitat.

When the first shipment of wolves arrived in the park and placed in holding pens, one of the neighboring tribes to the Bison Peak pack—the Crystal Bench pack—yipped and barked as soon as they caught wind of the curiously wafting scents. Within days, the coyotes had encircled the pens and come face to face with the new dogs on the block. At another pen in Rose Creek, within the territorial boundary of the Bison Peak pack, the entire pack vocalized and barked for nearly 30 minutes. They asserted themselves by angrily yip-howling. "I don't know what they were saying, but I'm sure they were not happy about the new arrivals," Crabtree said. "Things are going to change and coyotes on the northern range of Yellowstone are about to be usurped of their status in the park. The wolf has come back to claim its old niche."

The transition could be long and gradual or it might be short and violent; but it won't happen overnight, because for the rest of the decade, at least, and until wolves assert their own territorial dominance, coyotes will maintain a tenuous foothold. And then, as wolves take center stage, coyotes will thin out, their tracks will fade. It is important to remember that before the wolves were extirpated, we had a diverse and healthy enough ecosystem to allow another large canid to take the wolf's place and help keep the balance of Yellowstone going. Coyotes are the survivors, the great predator that humans couldn't extirpate. In the Lamar Valley they will now be bowing out the canid way.

Perhaps the only human record of the Bison Peak pack and its empire will be Bob Crabtree's research. Who will remember the song of the coyote in the face of wolves? To answer this question, we need to consider a paradox. Wolves may be the very reason why coyotes are irrepressible and were able to replace them. "Why do they seem to do so well from exploitation?" Crabtree asks. "Maybe we have the wolf to thank or blame for that. Wouldn't it be ironic if wolves did indeed descend from coyotes rather than the reverse. I find a purposefulness in that because it is almost like a Native American legend telling us: The coyote created the wolf to make itself stronger not only against wolves but to cope with the ultimate predator: Those on two legs."

The last page of the Lamar Pack story has yet to be written. For now the coyotes claim their territory as they have throughout the ages.

Picture yourself sitting around a campfire under the firmament of the stars. Suddenly, you hear a rustling in the brush and then, silhouetted against the moon, is the shape of four canids. Raising their muzzles to the world above, their group howling sends a chill down your spine.

How many coyotes are out there right now yipping to the dusk skies of North America? An estimate would be in the millions but still it is pure speculation. Coyotes are writing a new chapter in their natural history. In this case, with this remarkable animal, numbers appear to be inconsequential because the coyote is so smart and malleable it adjusts the size of its clan to fit the changing landscape. Where the environment is stable and healthy, coyotes bestow family values, espouse social bonds and group in packs just like wolves. Where war has been declared on its numbers the coyote has shown us its mettle as a survivor.

Humans have yet to learn the lesson of how to live in accordance with the land. Coyote, as Native American legend suggests, is there to teach us valuable lessons. But are we willing to listen? Are we capable of recognizing coyote's beauty, its rapturous music and its clever mind? The more we understand about coyote, it seems, the more remarkable the animal becomes.

Still, the fate of the coyote could be a reflection of our own future. One Native American story says that in the end of time, when humans and other life forms are gone, the coyote will remain, outlasting us all. But at the same time, the legend adds, the coyote does not like to be left alone, and prefers to be social.

And now we have come to a point of decision. We can either walk together on the path ahead or we can choose to go our separate ways. The former gives us a singing friend to send shivers of wonderment down our spine; the latter may lead to a very lonely world. Either way, the coyote will be there in the cooling moments of dawn, announcing the start of another new day.

A transient coyote uses its keen sense of smell to detect the individual odors of pack members in a foreign territory. Based upon what its nose tells it about the local occupants, it then will decide whether to intrude or retreat.

How to Save a Coyote

In the past, anti-coyote passions have been a way of life for some sheep ranchers, but today a foresighted non-profit organization is seeking a cease-fire by appealing to wool growers in a place where they are willing to listen—their wallets. "We discovered the best way to change the way ranchers treat coyotes is to appeal to their economic interests and provide an incentive to be tolerant. To think before pulling the trigger," says Lill Erickson, a conservationist from Livingston, Montana who together with Dude Tyler and a diverse range of citizens founded "Predator Friendly Wool."

Although it sounds like an oxymoron, Predator Friendly Wool was born with the idea that protecting predators and producing wool need not be mutually exclusive. The organization pays ranchers a premium price for wool if they agree to raise their sheep, and possibly absorb small losses, without killing coyotes and other predators. By purchasing a wool Predator Friendly coat, hat, or mittens, buyers are helping coyotes.

Free-market environmentalism is regarded as an emerging alternative to what some perceive as heavy handed laws mandating protection of species on private land, says Terry Anderson, a natural resource economist with the Political Economy Research Center—a conservative think-tank that examines natural resource issues. "Anything that makes predators more valuable to the land owners is a positive step toward saving them. The key is to make predators an asset instead of a liability."

Readers who want more information about how they can purchase Predator Friendly products should write to Predator Friendly Wool, 1300 Springhill Road, Belgrade, Montana 59714.

These groups can be contacted for more information about coyotes and other predators:

Yellowstone Ecosystem Studies
P.O. Box 6640
Bozeman, MT 59771

National Wildlife Federation
1400 16th Street NW
Washington D.C. 20036

Defenders of Wildlife
1244 19th Street NW
Washington D.C. 20036

Predator Project
P.O. Box 6733
Bozeman, MT 59771

National Audubon Society
801 Pennsylvania Avenue SE
Washington D.C. 20036

Wildlife Damage Review
P.O. Box 85218
Tucson, AZ 85754